THE THEORY OF PHYSICS

A Comprehensive Guide to the Concepts of Physics

Volume IV: Quantum Mechanics

Noah M. MacKay

based on the university notes made by
Noah M. MacKay, M.Sc.

whose notes were taken in the undergraduate and graduate courses led by
Dr. Gregory Lapicki & Dr. Michael Dingfelder

First Edition, 2020

Second Edition, 2022

Contents

1 Notes from the Author — 7

2 Introduction: The Quantum World — 9

I Mathematical Introduction — 13

3 The Mathematics of Quantum Mechanics — 15
   3.1 Probability — 16
   3.2 Operators — 17

II Basic Quantum Mechanics — 19

4 The Schrödinger Equation — 21
   4.1 The Quantum Hamiltonian — 22
   4.2 Free Particles — 23
   4.3 Particles in an Infinite Well — 25
       4.3.1 Particles in a Box Potential Well — 27
   4.4 Particles in a Linear Potential Well — 29

5 Simple Harmonic Oscillator — 33
   5.1 1-D Spatial Basis — 34
   5.2 Energy Basis — 36

III Rotational Quantum Mechanics — 39

6 Angular Momentum — 41
   6.1 Azimuthal Momentum — 42
   6.2 Intrinsic Momentum - Spin — 44

	6.3	Total Angular Momentum	46

7 Rotational Hamiltonian — 49
  7.1 The Hydrogen Atom 50
  7.2 Spin-dependent Hydrogen Atom 52

IV Approximations — 55

8 Variational Method — 57
  8.1 Binding Energy of Helium 59

9 Perturbation Theory — 61
  9.1 General Theory 62
  9.2 Free Particle Scattering 64
  9.3 Electron Repulsion 65

V Relativistic Quantum Mechanics — 67

10 Special Relativity and Quantum Mechanics — 69
  10.1 Klein-Gordon Equation 70
  10.2 Dirac Equation 71
  10.3 A Brief Intro to Quantum Field Theory 73

11 Gravity and Quantum Mechanics — 75
  11.1 String Theory 76
    11.1.1 Particles as Strings 77
    11.1.2 Dimensions as Membranes 79
  11.2 Loop Quantum Gravity 81
    11.2.1 The Loops in LQG 83
  11.3 Quantum Particle of Gravity 84
    11.3.1 Spin of a Graviton 85
    11.3.2 Gravitational Wave-Particle Duality 86

VI Problems — 89

12 Exercise Problems — 91

VII Acknowledgements 97

13 About the Author 99

14 Thanks 101

Chapter 1

Notes from the Author

To those who bought, read, enjoyed and learned from the first three volumes of *The Theory of Physics*; I thank you from the bottom of my heart. The goal of these books is to share the passion and love I have towards physics, and to teach you what I've learned in university. These books are not, by any means, a replacement to a physics lecture, but rather its companion.

To reintroduce myself, my name is Noah Matthew MacKay. I wrote *The Theory of Physics, Volumes 1 & 2* initially out of frustration towards the way certain professors taught physics. Introductory physics, or as I would like to call jokingly "baby physics," evolves around elementary classical physics and electromagnetism: the very topics of the first two books. When I wrote the original first addition, I was a graduate student looking in retrospect of my undergraduate studies. Now writing this second edition, I am a PhD student looking in retrospect of my graduate studies, how the things I learned can be beneficial in revising this book.

I wrote the first volume based on my university notes on calculus-based physics and classical mechanics, already knowing I would have to write the second book on electromagnetism (also based on my notes), and perhaps a few more books afterwards. As I wrote the third volume on select topics of modern physics, it was tremendous fun writing about the most intricate laws of science, and how seemingly out-of-touch advanced physics are in connection to the "golden oldie" classical laws. Now, I present to you the fourth book based on quantum mechanics, which (like modern physics) spans from the 1900's to today. Of course, this means physics currently in progress or evolving as of today or in the near future. This is where introductory physics comes to an end, as we step into the realm of advanced physics. But no fear, for I am here.

My passion for physics started when I was a young boy, and it lived on with me into university, when I attended East Carolina University to study physics along with mathematics and German language.

There at East Carolina University, I saw a physics department who is caring towards their students and is as passionate towards physics as I am. However,there were professors who had notoriously yet unintentionally discouraged their students from continuing their physics studies. When I tutored students who had those professors, they would always tell me how I made something that was made harder much easier to comprehend. This is why I am writing these books.

The two previous books, this current book, and the future books that would follow, are meant to introduce various fields of physics with simple language and a good basic understanding thereof, no matter how complex they may seem at first. While doing so, I want to perserve and showcase the elegance of the mathematical equations to these scientific concepts.

In this book, I am not beating around the bush when I talk physics. The way I see it, as long as I am here guiding the way and explaining the in's and out's of a certain concept in physics, there shouldn't be a reason to be afraid of what the math looks like.

I approach each topic with a goal to set a mood of eagerness towards that topic. By giving a brief history thereof; giving basic, real-world examples and a straight-forward, no-bullshit definition; I want to make it clear that the topic in question is relevant and an essential part of everyday life. And to that extent, I want to prove that physics isn't a dead science, but a lively one whose goal is to solve this jigsaw puzzle that is our understanding of the universe.

Chapter 2

Introduction: The Quantum World

In the past books, we discussed the physics of matter and waves seperately, especially in Vol. I on classical mechanics. However, Vol. III discussed on the subject of "wave-particle duality," beginning with Planck's notion of energy quanta and ending with Bohr's model of the hydrogen atom. As we now consider quantum mechanics, we must first understand what sets the quantum world apart from the classical world.

There was a lecture that the physicist Richard Feynman led at Cornell University. There he lectured about the dueling nature of particles and waves, how they can coexist as a "quantum mechanical wave." Using the analogy of particles to bullets and waves to the typical water / sound waves, Feynman introduces wave-particle duality with the ancient argument of light being a wave or a particle (an argument we recall from Vol. II). He mentions that electrons, first introduced into the world of science as particles, are also seen in quantum mechanics as fragments of a probability fog surrounding an atom. In quantum mechanics, matter and radiation are acknowledged as one in the same, unlike in classical mechanics.

As we came across the *theoretical* origins of wave-particle duality with Planck's notion, Richard Feynman begins the lecture with the *experimental* origins of wave-particle duality: the double slit experiment. Starting with his particle "bullets," Feynman explains that a source releases "clumps of bullets" - packets of particles - which impact an observation panel displaced from a board cut with two evenly-spaced-apart slits. The probability density of the particles on the observation panel, N_{12}, is but the sum of the two individual probability densities N_1 and N_2, one for each slit.

CHAPTER 2. INTRODUCTION: THE QUANTUM WORLD

Figure 2.1: Richard Feynman.

Once the experiment is repeated with waves, the probability density of the wave on the observation panel is instead a wave interference pattern. This contradicts the superposition notion of particles, for the intensity of the wave interference is not a direct sum of the two individual waves. However, wave interference involves the superposition of the interacting amplitudes. The intensity, in this case, is proportional to the square of the superposition, $N_{12} = |A_{12}|^2$.

After discussing what was expected from the Double Slit Experiment, we now experiment with the electrons. Electrons were first discovered as particles, so it is expected that they behave as such. Meaning, their probability density should result in a superposition. Nonetheless, the resulting probability density function is that of a wave interference! Should either of the double slits be covered, the electrons probability curve mimics that of particles (individually as N_1 and N_2), but with both slits open it is a wave interference.

Repeating the experiment with electrons again, but this time with a light source emitting immense light between the slits on the outside, only then the probability density curve of electrons looks like that of classical particles. With no light source, the probability curve is a wave interference. With a light source, the curve is a superposition. This raises another question: how does the light interact with the electrons?

The interaction between electrons and light resurrects the ancient argument of light being a wave or a particle. For electrons to be influenced by light, light must be made of particles for these electrons to interact with and be scattered by. This led to the theoretical understanding of light containing quanta called photons. The absence or the dimming of the light source is the absence or decrease of photon count, which reverts the superposition curve back into a wave interference curve. This provides the basis of wave-particle duality for both light and the electrons (experimentally).

In either case that the light source is on or off during the experiment, the confusion of the result remains in the question: does the individual electron go into slit 1 or slit 2, or even both? Feynman recounts Heisenberg's response to the puzzling question, which is the Uncertainty Principle,

$$\Delta p \Delta x \geq \frac{\hbar}{2}.$$

No matter how delicate the experiment may be, it is impossible to know from which slit an individual electron passes through without disrupting the electron, itself. In extension to the experiment's own Uncertainty Principle, there is no way to predict ahead of time through which slit an electron may pass. Only upon observation of the electron during the experiment, either with the light on or off, will you know. As Feynman commented, because there is no prior prediction to such a phenomenon, physics has given up its initial purpose of predicting the future, and has therefore resorted to the theory of probability.

Part I

Mathematical Introduction

Chapter 3

The Mathematics of Quantum Mechanics

Just as Richard Feynman concluded in his lecture on quantum physics, physics has given up on being predictable. Especially with objects we cannot see, yet acknowledge to be true and existent, all we can say is that either it is there or not... or that we don't know for sure. For instance, we acknowledge photons exist, they are the particle counterpart of light waves. We know how fast they are going (it's the speed of light c), but yet we can't say for sure where a photon may be in a single ray of light.

Similarly for breathable air, we know based on chemistry that air is made of nitrogen, oxygen, argon, and other molecules. We take an empty box, which is full of air, and close the lid. There is 100% certainty that there is air in the box. But for a single nitrogen, oxygen or argon molecule, I do not know where they are precisely inside my box.

Consider also looking at very fast, uniform waves with an acknowledged constant wavelength and frequency (such as light). As we let the wave pass by, all we can possibly measure is its speed and kinetic energy (just like how the LIGO detectors measured gravity waves). However, if we want to know the wave's wavelength and frequency instead, it is required that we freeze a time frame (so time is now known) and measure the wavelength. But with the frame frozen, it would be hard to tell whether the wave is really stationary or in motion. With a large uncertainty in its speed with a known wavelength and time frame, how is it possible to find the frequency of the wave?

3.1 Probability

Probability is what I call mathematical cluelessness. It is a value of certainty between 0 (uncertainty) and 1 (complete certainty). For subatomic particles we cannot see, but know to exist, their quantum mechanics is dictated by a value of probability: an integer between 0 and 1.

For matter and radiation are interchangeable, consider a wave function $\psi(x)$. As a function of a radiation wave, it is also the materialistic entity of subatomic matter. According to Feynman, the probability density for a wave is the square of a composition of wave amplitudes: $\text{Pr} = |A|^2$.

Letting waves extend all across space, the wave's amplitude is a specific number that determines its height / deepness from equilibrium. In order to determine this number, one must integrate the function across the entire region, letting it be within a certain range or from negative infinity to positive infinity. So, since

$$|A| = \int |\psi(x)| dx$$

(and squaring the amplitude means squaring the wave function), then the probability density of a particular wave function is

$$\text{Pr} = \int |\psi(x)|^2 dx$$

However, if $\psi(x)$ is a complex function, such as $\psi(x) = e^{ikx}$, then $|\psi(x)|^2 = \psi^*(x)\psi(x)$, where $\psi^*(x)$ is the conjugate of a complex function (such as $\psi^*(x) = e^{-ikx}$).

Keeping with probability-based physics, to say that a particle is within a specific region (no matter where it may be precisely), the probability density is equal to 1. This is called "normalization:"

$$1 = \int \psi^*(x)\psi(x) dx$$

which is crucial in determining what a specific wave function is based on conditions and assumptions.

It is important that every quantum wave function is normalized, for the probability density of a particle's existence is indeed 100%, given it is present in all space.

3.2 Operators

Because quantum physics is based on probability, mathematical exactness regarding a particle's particular property is replaced by mathematical "what-if's." According to the Heisenberg uncertainty principle, derived by German physicist Werner Heisenberg in 1927,

$$\Delta p \Delta x \geq \frac{\hbar}{2}$$

the precision of an object's location creates a shroud of uncertainty around its speed (or momentum), and vice versa.

Letting the position x be the variable, let us isolate the uncertainty of momentum:

$$\Delta p = \frac{\hbar}{2\Delta x}$$

Because we chose x to be our variable, we can allow its range of certainty Δx to be precise or shrouded. However, the precision of position inversely affects the precision of momentum Δp.

To make momentum precise, in order to define quantum momentum, we must make Δx to be very broad - so broad, it is paired with the imaginary number i. Luckily, while the position uncertainty is in the denominator, a wide change of displacement becomes a finite increment of $d/(i\,dx)$, or $-i\nabla = -i\partial_x$.

Therefore, we have defined momentum as a differential operator:

$$\widehat{p} = -i\hbar\nabla$$

given that the position operator is just the independent variable,

$$\widehat{x} = x$$

Also given that momentum is $p = \hbar k$ based on the de Broglie wavelength, $\widehat{k} = -i\nabla$ is the wave number operator.

Now consider the time-dependent version of Heisenberg's uncertainty principle:

$$\Delta E \Delta t \geq \frac{\hbar}{2}.$$

Here, the precision of energy and time change inversely with each other, just like momentum and position. Using the same methodology used to derive the momentum operator, we can define the energy operator:

$$\widehat{E} = i\hbar \partial_t$$

given that the time operator is just the independent variable,

$$\widehat{t} = t$$

Also given that energy is $E = \hbar \omega$ based on the Compton wavelength, $\widehat{\omega} = i\partial_t$ is the frequency operator.

Every operator has a corresponding "eigenvalue:" a particular value that determines the operator in question. For instance: $\widehat{p}\psi = \hbar k \psi$, and $\widehat{E}\psi = \hbar \omega \psi$. Specific energy E can also be an eigenvalue, itself, which is the eigenvalue of the "Hamiltonian operator:" $\widehat{H}\psi = E\psi$. This is called the Schrödinger equation.

Part II

Basic Quantum Mechanics

Chapter 4

The Schrödinger Equation

In 1833, long before the conception of relativity or quantum physics, Irish mathematician William Rowan Hamilton postulated a particular form of classical mechanics that not only satisfied Newton's laws, but also looked at energy as momentum-dependent. Hamilton would suppose the "Hamiltonian formalism," which states that the total energy of a single particle ultimately depends on its momentum and its potential energy:

$$E = \frac{p^2}{2m} + V \qquad (4.1)$$

where the first term is the kinetic energy and the second term is the potential energy. Note that the right-hand side of this equation, itself, is called the *Hamiltonian*, labeled as H.

The competitor to Hamilton's framework of particle energy was the one belonging to Italian-born French mathematician Joseph-Louis Lagrange, who thought that the energy of a stochastic, randomly moving particle must be mathematically reduced to find its exact physics. Instead of adding kinetic energy with potential, Lagrange subtracted the two energies, restraining the particle to have least action.

But as the framework of particle mechanics shifted from classical physics to quantum physics, applying the wave-particle duality would mean we would look at particles as waves. Meaning, exactness is replaced by probability. In 1926, Austrian physicist Erwin Schrödinger had considered the developments of wave-particle duality, and thus used the Hamiltonian as a way to describe the quantum conservation of energy.

Figure 4.1: Erwin Schrödinger.

4.1 The Quantum Hamiltonian

Looking at the Hamiltonian formalism for total energy, both momentum and energy must imply Heisenberg's uncertainty principle. Given a quantum mechanical wave with a function ψ, momentum and energy must be described by their operators:

$$\widehat{E}\psi = \frac{\widehat{p}^2}{2m}\psi + V(x)\psi$$

letting the potential energy be a function of displacement.

Explicitly writing out the operators, the quantized Hamiltonian is defined as follows:

$$\boxed{i\hbar\partial_t\psi = \frac{-\hbar^2}{2m}\nabla^2\psi + V(x)\psi} \qquad (4.2)$$

which is called the Schrödinger equation, which is an essential equation in describing quantum phenomena.

In the Schrödinger equation, we have the Hamiltonian operator:

$$\widehat{H}\psi = \frac{-\hbar^2}{2m}\nabla^2\psi + V(x)\psi$$

where the first term is the kinetic energy operator and the second term, obviously, the potential energy; as well as how the Hamiltonian is defined. If we consider an interaction that is time-dependent,

$$\widehat{H}\psi = i\hbar\partial_t\psi.$$

However, if the interaction is time-independent, the energy operator is replaced by an energy "eigenvalue:"

$$\widehat{H}\psi = E\psi.$$

When Schrödinger announced his equation to the physics community, he won the Nobel physics prize in 1933. However, despite drafting an equation that explains all of quantum physics, Schrödinger did not know exactly what this quantum wave function even represented.

The role of the wave function was unknown, until German Max Born determined that the wave function depicts the probability of existance for a particular particle. Wherever the peaks of the wave function are prominent, that is where a particle is located with the highest probability, however not an integer close to 1.

4.2 Free Particles

Free particles are considered to be kinetic particles unbound to any potential energy $(V(x) = 0)$. These particles are, namely, free as they move across

all space. Considering the time-independent Schrödinger equation with no potential,
$$\frac{-\hbar^2}{2m}\nabla^2\psi = E\psi,$$
where $\psi = \psi(x)$ is a wave function of variable x. Considering this is a second-order differential equation, where the derivatives of the function produce the original function, let us have the ansatz $\psi = \exp(ax)$. Therefore,
$$\frac{-\hbar^2}{2m}a^2 = E$$
$$\Rightarrow a = \sqrt{\frac{2mE}{-\hbar^2}} = \frac{i}{\hbar}\sqrt{2mE} \qquad (4.3)$$
which sets the wave function equal to
$$\psi(x) = \exp\left[\frac{ix}{\hbar}\sqrt{2mE}\right] = A\sin\left(\frac{x}{\hbar}\sqrt{2mE}\right) + B\cos\left(\frac{x}{\hbar}\sqrt{2mE}\right) \qquad (4.4)$$

For particles in boundless free space, Eq. (4.4) would be the wave function of any generic particle. To make the function cleaner, suppose the particle is in a region of all space (from negative infinity to positive infinity).

At the hypothetical boundaries of the two infinities, the probability of the particle existing there would be near zero. Sure, the particle exists along my region with 100% certainty, but how certain am I that the particle in question is even there at the boundaries of infinity? I'm not.

How certain am I that a Ukrainian (which there are a lot of them) is taking a flight to Berlin? I cannot say for sure without any observation. And say I'm located in Raleigh, NC; I cannot observe a Ukrainian taking a flight to Berlin. That boundary of infinity is unknown to me, so I reduce my probability to near zero (without jumping to the conclusion it is zero, because that event is likely to happen even despite the lack of observation).

At the boundaries of infinity, because I am unsure of the particle's status there, I reduce its energy to zero (how do I know it is moving at all?). This makes the sine function equal to zero, and my cosine equal to 1. But since we agree that our probability at the boundaries of infinity are zero, let $B = 0$. So, the wave function of a free particle in all space is
$$\psi(x) = A\sin\left(\frac{x}{\hbar}\sqrt{2mE}\right) \qquad (4.5)$$

4.3. PARTICLES IN AN INFINITE WELL

The coefficent A is determined by the boundary conditions, which is solved by normalization. For all space, where the region is boundless, the coefficient is dependent on particle's energy and mass rather than where it is located within the region. So, with

$$A = \sqrt{\frac{2\sqrt{2mE}}{\hbar\pi}} \Rightarrow \sqrt{\frac{2mv}{\hbar\pi}}$$

where v is the speed of the free particle,

$$\boxed{\psi(x) = \sqrt{\frac{2mv}{\hbar\pi}} \sin\left(\frac{mv}{\hbar}x\right)} \tag{4.6}$$

4.3 Particles in an Infinite Well

Particles in an infinite well are still free particles, but they are confined to a region of space much shorter in length compared to all space. And since the well is "infinite," there is no way for the particles to escape, for $V(x) = \infty$ outside the well.

Imagine a water well layered with brick and mortar, the depth of the well is defined by how many layers of brick were stacked to form the well. Much like the water well, the infinite well for free particles has layers. Instead of stacked bricks, the depth of this infinite well is defined by "energy states." We discussed energy states in the previous book, when we mentioned Bohr's model of the hydrogen atom. The principle still applies, although an infinite well is nowhere near related to the hydrogen atom.

The Schrödinger equation is still the same as before, and so the solution to the wave function is nonetheless Eq. (4.5). But, given the region of space is within the boundaries of the well, we much change our system's boundary conditions, and the function itself.

Consider the revision of Eq. (4.5) as the following:

$$\psi(x) = A \sin(kx)$$

where $k = \sqrt{2mE}/\hbar$. Just like we considered with the boundaries of infinity, the probability in which our welled particle is located at the boundaries is near zero, It is more likely it'd be lounging around in the middle of the well than it would along the walls.

Since our wave function is also a sine function, the sine is zero at angles $0, \pi, 2\pi, \ldots$. In other words, all whole integers of π is where the sine function is zero. Also, we've decided that our probability would be zero along the walls of the well (let's say from $x = -L$ to $x = L$, making $2L$ the region of our well).

Therefore, making k be related to the whole integers of π and the length L:
$$k = \frac{n\pi}{L}$$
our wave function is now
$$\psi(x) = A \sin\left(\frac{n\pi}{L}x\right)$$

Since k has two definitions, it is possible to define what the energy of the system E would be in an infinite well:
$$E = \frac{n^2\pi^2\hbar^2}{2mL^2} \tag{4.7}$$

As we go about normalizing the wave function, where the region spans from $-L$ to L, the coefficient A becomes $\sqrt{2/L}$. This finalizes the wave function in an infinite well to be

$$\boxed{\psi(x) = \sqrt{\frac{2}{L}} \sin\left(\frac{n\pi}{L}x\right)} \tag{4.8}$$

which applies to any well of length L, provided that $V(x) = 0$ inside and $V(x) = \infty$ outside.

These whole integers of n are the energy states. As n increases, the sine function becomes more chaotic due to the increased energy of the particle. Given in the energy equation and in the sine function, the ground state of a particle in any well is $n = 1$, which is where the particle would have the lowest non-zero energy.

If a particle is present in a well, it cannot reside in a "null state" of $n = 0$, because it is simply not possible. To suggest $n = 0$, you're suggesting the particle does not exist in the well, or it isn't there. But since it is, it must have some base energy, which is at the ground state of $n = 1$.

4.3. PARTICLES IN AN INFINITE WELL

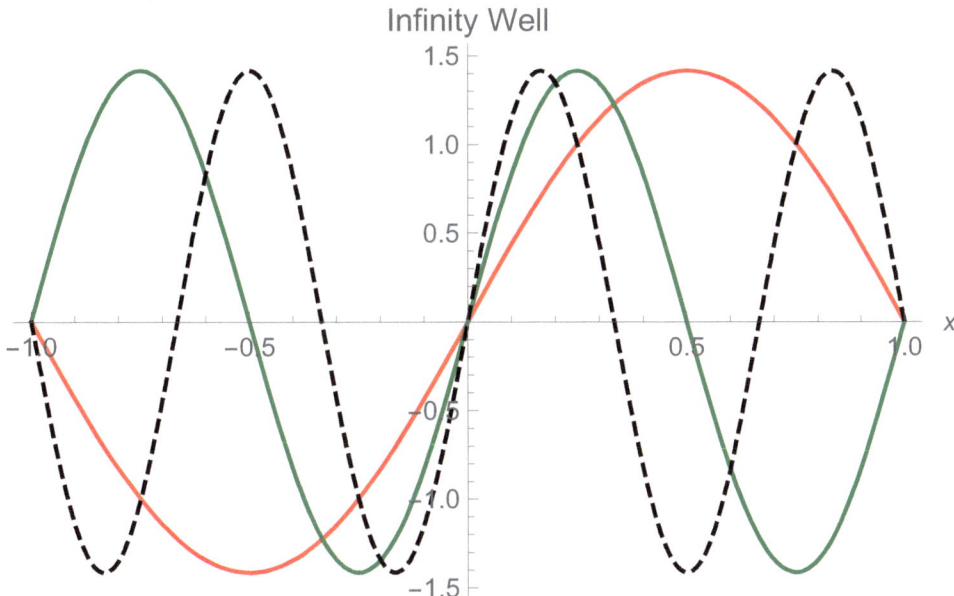

Figure 4.2: The infinite well for $L = 1$ and states $n = 1$ (red), 2 (green) and 3 (black dashed).

4.3.1 Particles in a Box Potential Well

In this system, a particle is inside a well defined by an arbitrary potential $V(x) = V$. Note that inside the potential well, the particles are free. Yet, these particles can *penetrate* through the walls to escape and become free partcles outside the well. This is called "tunneling."

In this scenario, the time-independent Schrödinger equation looks quite different:
$$\frac{-\hbar^2}{2m}\nabla^2\psi = (E - V)\psi,$$
let us call $E - V = M$, some energy value. Because V is defined as arbitrary, it can either be greather than, less than, or equal to the energy E.

Considering the ansatz $\psi = \exp(ax)$,
$$-a^2 = M\frac{2m}{\hbar^2}$$
which depends on the status of $M = E - V$. If $V > E$, so the particle is tunneling through the walls of the potential well, then M is negative:
$$a = \pm\sqrt{\frac{2m(E - V)}{\hbar^2}} = \pm k$$

If $V = E$, so the tunnelling particle has surfaced from the potential well, then $M = 0$, making $a = 0$. If $V < E$, so the particle has escaped and it is a free particle in all space, then M is positive:

$$a = i\sqrt{\frac{2m(E-V)}{\hbar^2}} = i\kappa$$

Note that $k \neq \kappa$, for they both depend on the energy E in comparison to the potential V.

We have already come across the function for a free particle in all space:

$$\psi(x) = \exp[i\kappa x] = \sqrt{\frac{2mv}{\hbar\pi}} \sin\left(\frac{mv}{\hbar}x\right) \qquad (4.9)$$

and the wave function of $a = 0$ is a constant value $\psi = 1$.

For a particle that is tunneling through the potential barriers of the well, the wave function is

$$\boxed{\psi(x) = \exp[\pm kx] = A_0 \exp\left[\pm\sqrt{\frac{2m(E-V)}{\hbar^2}}\,x\right]} \qquad (4.10)$$

depending on which end a particle may tunnel through to escape.

The probability of tunneling is defined as

$$\boxed{\text{T} = A_0^2 \int \exp\left[\pm 2\sqrt{\frac{2m(E-V)}{\hbar^2}}\,x\right] dx} \qquad (4.11)$$

If the probability is 1 (meaning the likelihood of tunelling success is very certain), then

$$A_0 = \sqrt{2k}$$

However, if the probability of tunneling is zero (meaning the particle is reflected and richochetted back and forth like a ping pong ball), then $A_0 = 0$.

Otherwise, the probability of tunneling and the probability of reflection are added to become 1 (the probability of a particle doing *something* is 1):

$$\text{T} + \text{R} = 1 \quad \text{or} \quad \text{R} = 1 - \text{T} \qquad (4.12)$$

4.4. PARTICLES IN A LINEAR POTENTIAL WELL

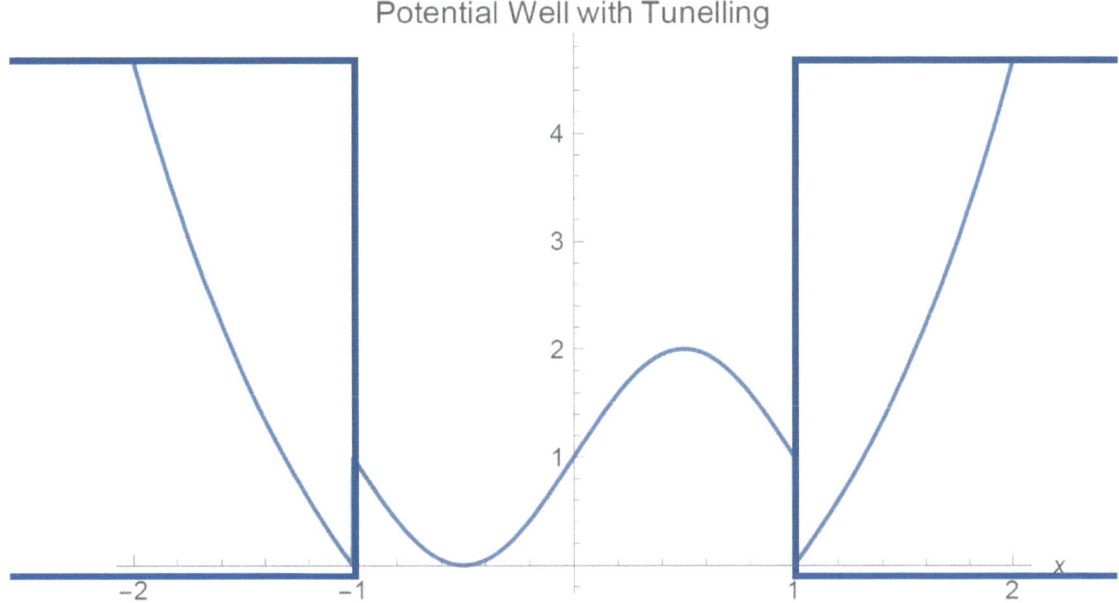

Figure 4.3: The potential well with boundaries $L = -1$ and $L = 1$ at state $n = 1$.

4.4 Particles in a Linear Potential Well

In this system, a particle is inside a symmetric linear potential well with the profile $V(x) = F|x|$. The force $F(x) = -\partial_x V(x)$ is a constant, which pays homage to introductory physics with constant forces (such as $F = ma$) and linear potentials (such as $U = Fx$). If the position x is instead used as off-equilibrium displacements, akin to Hooke's law, the linear potential well is applied to complex systems known to exhibit "self-organizing criticality," or a moment where instable systems regulate themselves with a chaotic avalanche.

The corresponding time-independent Schrödinger equation is

$$\frac{-\hbar^2}{2m}\nabla^2\psi(x) + F|x|\psi(x) = E\psi(x)$$

To clean up the equation, we introduce a length scale of $x = au$, where a is length and u is unitless. This turns the Schrödinger equation into

$$-\psi''(u) + 2\left(\frac{Fma^3}{\hbar^2}|u| - \frac{Ema^2}{\hbar^2}\right)\psi(u) = 0$$

To make the factor in front of $|u|$ be equal to one, we define $a^3 = \hbar^2/(Fm)$ so that the length scaling is

$$a = \left(\frac{\hbar^2}{Fm}\right)^{1/3} \tag{4.13}$$

We also define a dimensionless parameter $Ema^2/\hbar^2 = \epsilon$, and therewith the variable $\xi = |u| - \epsilon$. This solves for the wave function as follows:

$$\psi(\xi) = A_0 \mathrm{Ai}\left(2^{1/3}\xi\right) + B_0 \mathrm{Bi}\left(2^{1/3}\xi\right)$$

Here, $\mathrm{Ai}(y), \mathrm{Bi}(y)$ are the so-called Airy functions: $\mathrm{Ai}(y)$ converges to zero for positive y and $\mathrm{Bi}(y)$ diverges for positive y; both are sine and cosine waves for negative y.

For a particle inside the linear well, where $|x|$ is always positive, it makes physical sense for the wavefunction to exist within the well and collapse outside it. Thus, $B_0 = 0$, and we only consider the convergent $\mathrm{Ai}(y)$ Airy function:

$$\psi(\xi) = A_0 \mathrm{Ai}\left(2^{1/3}\xi\right)$$

$$\rightarrow \boxed{\psi(x) = A_0 \mathrm{Ai}\left(2^{1/3}\frac{|x| - a\epsilon}{a}\right)} \tag{4.14}$$

Here, $a\epsilon = E/F$ denotes the characteristic displacement under which work is applied (i.e. $E = F \cdot x$). Normalization is evaluated over all space to obtain the coefficient A_0:

$$\boxed{A_0 = \left(2a\epsilon \mathrm{Ai}(-2^{1/3}\epsilon)^2 + 2^{2/3}a\mathrm{Ai}'(-2^{1/3}\epsilon)^2\right)^{-1/2}} \tag{4.15}$$

which depends on the corner node condition $\psi(0)$. For a system of particles, you would likely find particles resting at the corner node ($|x| = 0$) as a natural equilibrium point. As it is in nature, what comes up must come down. Thus, ϵ is the energy state n, as energy of the linear well system is

$$\boxed{E_n = n\left(\frac{\hbar^2 F^2}{m}\right)^{1/3}} \tag{4.16}$$

The displacement, $a\epsilon = E/F$, is based on quantum angular momentum:

$$\boxed{a\epsilon = x_n = n\hbar \left(\frac{\hbar F}{m}\right)^{1/3}} \tag{4.17}$$

4.4. PARTICLES IN A LINEAR POTENTIAL WELL

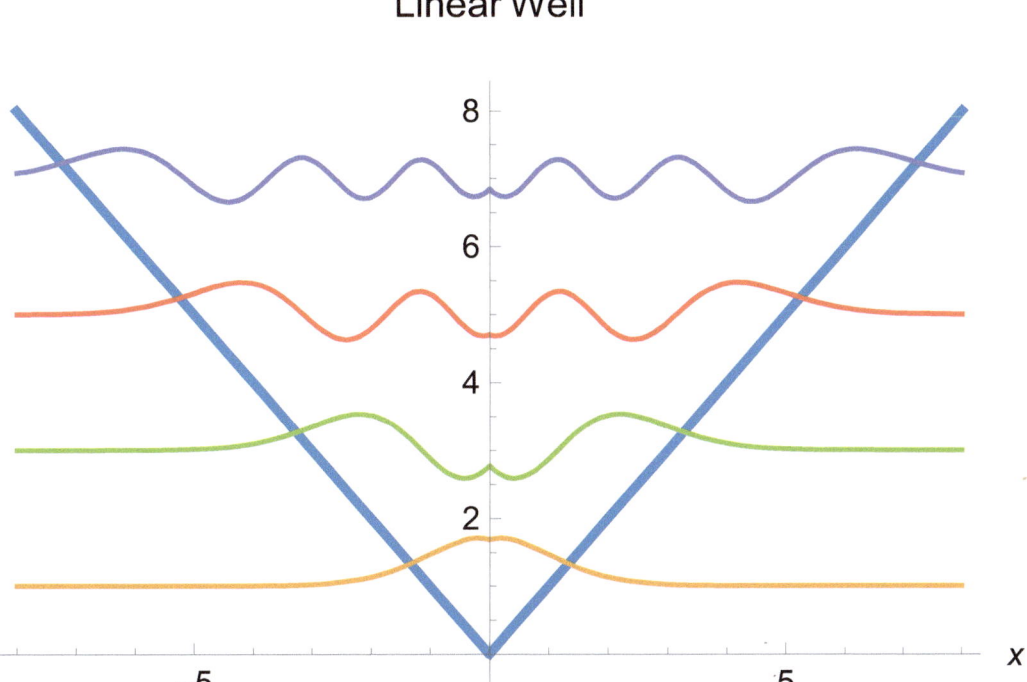

Figure 4.4: A symmetric linear potential well (profile in blue) with scaled force and mass $F, m = 1$. At the ground state $n = 1$ (in orange), the wave looks like a Gaussian bell curve for stationary particles. As the energy state increases (Green: $n = 3$, Red: $n = 5$, Purple: $n = 7$), the linear well wave function progressively behaves like a free wave. $\hbar = 1$ for scaling purposes.

Chapter 5

Simple Harmonic Oscillator

Recall from Vol. I that the potential energy of a spring oscillator is

$$U(x) = \frac{1}{2}m\omega^2 x^2$$

where m is the mass of the oscillator, ω is its angular frequency and x is the displacement from equilibrium.

In quantum mechanics, as we saw with the particle in the infinite well, a particle in a particular energy state has a corresponding value of energy. Provided that $E \propto n^2$, as n increases by an integer, the energy increases by its square. In other words, quantum particles have a tendency to escalate its energy.

For quantum particles in an infinite well, their escalated energy dictates the particles to do something, given a low probability to tunnel through an infinite well. So, they gyrate with translational, oscillating energy. Therefore, quantum particles behave like a harmonic oscillator with the potential of

$$V(x) = \frac{1}{2}m\,\widehat{\omega}^2\,\widehat{x}^2 \tag{5.1}$$

Letting our variable be x, the frequency operator is just the angular frequency ω. Placing Eq. (5.1) in place of $V(x)$ in the time-independent Schrödinger equation, we have the equation for a simple harmonic oscillator:

$$\boxed{\frac{-\hbar^2}{2m}\nabla^2\psi + \frac{1}{2}m\omega^2 x^2\psi = E\psi.} \tag{5.2}$$

5.1 1-D Spatial Basis

In this section, the Schrödinger equation for a simple harmonic oscillator is solved in one-dimensional space. Given Eq. (5.2), let us reorganize the differential equation to make it look cleaner:

$$\psi''(x) + \left(\frac{2mE}{\hbar^2} - \frac{m^2\omega^2}{\hbar^2}x^2\right)\psi(x) = 0$$

To make this equation easier to solve, let us rewrite x as a product of by, where b is a unit length and y is dimensionless:

$$\frac{1}{b^2}\psi''(y) + \left(\frac{2mE}{\hbar^2} - \frac{m^2\omega^2}{\hbar^2}b^2 y^2\right)\psi(y) = 0$$

$$\Rightarrow \psi''(y) + \left(\frac{2mEb^2}{\hbar^2} - \frac{m^2\omega^2}{\hbar^2}b^4 y^2\right)\psi(y) = 0$$

To make the coefficient for the y^2 term be 1, $b \equiv \sqrt{\hbar/m\omega}$. So, $mEb^2/\hbar^2 = E/\hbar\omega \equiv \varepsilon$, which is also dimensionless. Now, our differential equation is dimensionless, and much friendlier to solve:

$$\psi''(y) + (2\varepsilon - y^2)\psi(y) = 0 \tag{5.3}$$

Here, let us consider two directions, each of their results we will combine them in the end. Let's look at the cases where $\varepsilon \to 0$ and $y \to 0$. In the first case, the reduced equation comes out to be

$$\psi''(y) - y^2\psi(y) = 0$$

which solves for $\psi(y)$ to be

$$\psi(y) = Ay^m \exp\left(\frac{-y^2}{2}\right) \tag{5.4}$$

In the second case, the reduced equation comes out to be

$$\psi''(y) + 2\varepsilon\psi(y) = 0$$

which solves for $\psi(y)$ to be

$$\psi(y) = A\cos(\sqrt{2\varepsilon}y) = A + \sqrt{2\varepsilon}y + \ldots \tag{5.5}$$

where the latter is the cosine expanded as a Taylor expansion.

5.1. 1-D SPATIAL BASIS

Combining the two results forms a wave function of

$$\psi(y) = \left(Ay^m + B\sqrt{2\varepsilon}y^{m+1}\right)\exp\left(\frac{-y^2}{2}\right) \quad (5.6)$$

and changing $m = n - 2$ and the coefficients A and B based on the whole integer n,

$$\psi(y) = \sum_n C_n y^n \exp\left(\frac{-y^2}{2}\right)$$

which has the "recursion relation" (an equation to determine every other coefficient of C given an arbitrary coefficient C_n) of

$$\boxed{C_{n+2} = C_n \frac{2n + 1 - 2\varepsilon}{(n+2)(n+1)}} \quad (5.7)$$

which is based on dimensionless energy and the energy state.

In Eq. (5.7), C_{n+2} relates to the energy state two tiers higher or lower than the state occupied by a certain particle. Due to the improbability of quantum particles, there is no way to make a definite claim that the particle may transcend into any state higher or lower than one of the neighboring states.

While C_n is definite (it relates to the current occupied state), C_{n+2} is not, thus we make it equal to zero.

Now, $\varepsilon = E/\hbar\omega$ can be defined, which solves for the energy eigenvalue of a harmonic oscillator:

$$\varepsilon = \frac{2n+1}{2}$$

$$\boxed{\Rightarrow E = \hbar\omega\left(n + \frac{1}{2}\right)} \quad (5.8)$$

where, again, the lowest energy state of a harmonic oscillator is $n = 1$.

Letting $C_n y^n = H_n(y)$ be the "Hermite Polynomial," which relates to the transitioning of oscillator states, the final revision of the wave function for a harmonic oscillator $\psi(y) \to \psi(x)$ (after normalization) is

$$\boxed{\psi(x) = \left(\frac{m\omega}{2^{2n}\pi\hbar(n!)^2}\right)^{1/4}\exp\left(\frac{-m\omega x^2}{2\hbar}\right)H_n\left(\left(\frac{m\omega}{\hbar}\right)^{1/2}x\right)} \quad (5.9)$$

where, once again, $V(x) = m\omega^2 x^2/2$ and $b^2 = \hbar/m\omega$.

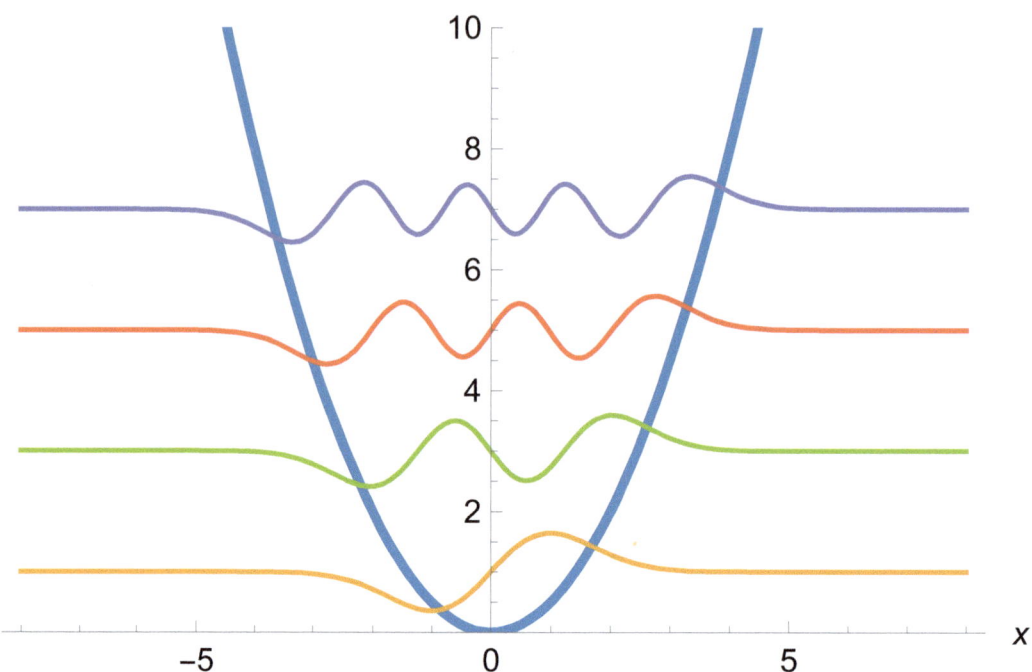

Figure 5.1: A simple harmonic oscillator (energy well drawn in blue) for states $n = 1$ (orange), 3 (green), 5 (red), and 7 (purple). As the states increase, the more obvious that particles exist as free waves inside the energy well, and flatten outside it.

5.2 Energy Basis

In this section, the Schrödinger equation for a simple harmonic oscillator is solved in the "energy space." This is meant to be an easier approach to the harmonic oscillator. Instead of looking at functions, we look at the energy states themselves, how a particle may ascend or descend into neighboring energy states.

In this scenerio, let us rewrite the one-dimensional wave function $\psi(x)$ as an energy function $|n\rangle$. What is consistent with the previous derivation of the harmonic oscillator is that the Hamiltonian eigenvalue is indeed $E = \hbar\omega(n + 1/2)$. So, in the energy space,

$$\widehat{H}_E|n\rangle = \hbar\omega\left(n + \frac{1}{2}\right)|n\rangle$$

5.2. ENERGY BASIS

$$\Rightarrow \frac{1}{\hbar\omega}\widehat{H}_E|n\rangle = \widetilde{H}_E|n\rangle = \varepsilon|n\rangle \tag{5.10}$$

where $\widetilde{H}_E$ is dimensionless and $\varepsilon = (n + 1/2)$.

Note that in the energy space, we are supposed to see the actual rising and lowering of a particle between energy states. So, let us rewrite the original Hamiltonian of the harmonic oscillator in terms of two essential operators: the "annihilation operator:"

$$\boxed{a \equiv \frac{1}{b\sqrt{2}}\widehat{x} + i\frac{b}{\hbar\sqrt{2}}\widehat{p}} \tag{5.11}$$

and the "creation operator:"

$$\boxed{a^\dagger \equiv \frac{1}{b\sqrt{2}}\widehat{x} - i\frac{b}{\hbar\sqrt{2}}\widehat{p}} \tag{5.12}$$

so that

$$\frac{-\hbar^2}{2m}\nabla^2 + \frac{1}{2}m\omega^2 x^2 \rightarrow \hbar\omega\left(a^\dagger a + \frac{1}{2}\right)$$

which verifies that $E = \hbar\omega(n + 1/2)$, provided that

$$\boxed{a^\dagger a|n\rangle = n|n\rangle} \tag{5.13}$$

In other words, if the annihilation operator involves descending states, and the creation operator involves ascending states, simultaneous descent and ascent makes the particle stationary in its currently-occupied state.

To prove that these new operators actually do what they are supposed to do, consider that a particle descending a state must "annihilate" itself in its original state in order to move down to one lower state:

$$a|n\rangle = C_n|n-1\rangle$$

and a particle ascending a state must "create" itself anew at the higher state in order to move up:

$$a^\dagger|n\rangle = C_{n+1}|n+1\rangle$$

Letting the energy function be complex, it shall have a conjugate $\langle n|$, where $(a|n\rangle)^\dagger = (C_n|n-1\rangle)^\dagger \Rightarrow \langle n|a^\dagger = \langle n-1|C_n^*$. And thus, by rule of normalization, $\langle n|n\rangle = 1$.

$$\langle n|a^\dagger a|n\rangle = C_n^* C_n \langle n-1|n-1\rangle = C_n^* C_n$$

$$\langle n|a^\dagger a|n\rangle = n\langle n|n\rangle = n$$

Therefore,
$$n = C_n^* C_n \tag{5.14}$$

and letting C_n be real (meaning its conjugate is itself),
$$C_n = \sqrt{n} \quad \text{and} \quad C_{n+1} = \sqrt{n+1} \tag{5.15}$$

Therefore, the annihilation and creation operators, defined by energy state, are
$$\boxed{a|n\rangle = \sqrt{n}\,|n-1\rangle} \tag{5.16}$$

and
$$\boxed{a^\dagger|n\rangle = \sqrt{n+1}\,|n+1\rangle} \tag{5.17}$$

which both do their jobs as intended.

Part III

Rotational Quantum Mechanics

Chapter 6

Angular Momentum

In classical mechanics, especially in Vol. I, we talked about the angular momentum of a rotating object being defined as the cross product between the radius of rotation and the object's tangential momentum:

$$\vec{L} = \vec{r} \times \vec{p}$$

What you may already know is that orbiting planets, such as the Earth, also have an intrinsic angular momentum $\vec{S}$, which constructs with the orbital angular momentum to form the *total* angular momentum:

$$\vec{J} = \vec{L} + \vec{S}$$

which either builds up, given the directions of L and S are parallel, or works against each other if their directions are anti-parallel.

As we zoom in on the quantum scale, nothing dramatic has changed in the knowledge of angular momentum. What we know of quantum particles already is that they are very energetic. In any energy well (either infinite or a standard potential well), these particles in energy states greater than $n = 1$ gyrate with translational energy, much like an oscillator (which we have just covered in the previous chapter).

From a side view, the oscillations look translational; they vibrate side to side, from one end of the well to another. However, assuming our well to be a parabolic salad bowl, the particle's gyrating instead looks like centripetal motion from the top view (each higher state increases the radius of rotation). Our quantum particles have angular momentum.

6.1 Azimuthal Momentum

Azimuthal angular momentum is the quantum equivalent to the earlier-mentioned classical angular momentum L. With our classical definition, it is possible to define quantum angular momentum as an operator, and thus we can figure our its eigenvalue to eventually put in the Schrödinger equation.

Considering two dimensions for r and p (so that we can apply polar coordinates for our centripetal motion), the angular momentum becomes

$$L_z = xp_y - yp_x$$

which is expanded in operator form as

$$\hat{L}_z = \widehat{xp_y} - \widehat{yp_x} = i\hbar(x\partial_y - y\partial_x)$$

Now transforming $(x, y) \to (r\cos\theta, r\sin\theta)$ (where r is a constant), the angular momentum is defined as

$$\hat{L}_z = i\hbar \left(\frac{\cos\theta}{\cos\theta}\partial_\theta + \frac{\sin\theta}{\sin\theta}\partial_\theta \right) = i\hbar\partial_\theta \qquad (6.1)$$

where a two is scaled away to avoid over-counting.

Eq. (6.1) is the angular momentum operator in our spatial frame. As we looked at harmonic oscillator in the "energy space," we can look at angular momentum in the "azimuthal space," where $\psi(x) \to |l\rangle$. Here, l is the "azimuthal number:"

$$\hat{L}_z|l\rangle = l_z|l\rangle$$

where l_z is its z-component, which influences the orientation of $\hat{L}$ along the z-axis.

Just like the energy state number n (which we saw in 1D space and energy space), the azimuthal number l is present in both azimuthal and 1D spaces. So,

$$i\hbar\partial_\theta \psi(r, \theta) = l_z \psi(r, \theta)$$

where the sample wave function here is solved to be

$$\psi(r, \theta) = R(r) \exp\left[i\frac{l_z \theta}{\hbar}\right]$$

One important thing to consider about centipetal motion is "azimuthal symmetry." This means that uniform circular rotation has no variation in its

6.1. AZIMUTHAL MOMENTUM

path based on angle projection. For instance, zero radians is also equal to 2π radians.

Letting $\theta = 0$ and 2π, based on azimuthal symmetry, $\psi(r, 0) = \psi(r, 2\pi)$:

$$1 = \exp\left[i\frac{2\pi l_z}{\hbar}\right]$$

which is Euler's Identity: $1 = \exp(i2\pi)$. To have Euler's identity be true, l_z must be a whole integer of $\hbar$, letting it be $m_l = 0, \pm 1, \pm 2, \ldots$. Therefore, we have defined l_z in our 1D space:

$$\boxed{\hat{L}_z \psi = m_l \hbar \psi} \qquad (6.2)$$

where our whole integer m_l is called the "magnetic quantum number." One key property of the magnetic number m_l is that it has $2l+1$ values, spanning from $-l, \ldots, 0, \ldots, l$.

For example, given $l = 1$, $m_l = 1, 0, -1$; which is a set of $2(1) + 1 = 3$ values.

The magnitude of the angular momentum can be derived through Eq. (6.2), where

$$L^2 = \hat{L} \cdot \hat{L} = \hbar^2 (m_l \cdot m_l)$$

With m_l having $2l + 1$ values, m_l^2 would roughly have $4l^2 + 2l$ values, letting $l > 1$. To avoid any over-counting, $m_l^2 \to l(l+1)$. Therefore

$$L^2 \psi = \hbar^2 l(l+1)\psi$$

$$\boxed{\Rightarrow |\hat{L}|\psi = \hbar\sqrt{l(l+1)}\psi} \qquad (6.3)$$

which is a positive value of $\hbar$.

Using the same example with $l = 1$ again, $|L| = \hbar\sqrt{1(1+1)} = \hbar\sqrt{2}$.

To determine l based on the energy state n, let $l = n - 1$. Therefore, with the energy state at ground, i.e. $n = 1$, then $l = 0$. Meaning, the ground state is at a minimal energy that is too weak to have any angular momentum. But, at one state higher ($n = 2$), we have $l = 1$, which gives us angular momentum based on quantum gyrating.

6.2 Intrinsic Momentum - Spin

Quantum intrinsic momentum S is also called *spin*. It is a unique property of quantum particles that has a bit of a "twist." For physics purposes, spin may be thought as the intrinsic rotations of a particle. However, in actuality, spin has nothing to do with "spinning." It is purely quantum-mechanical, for it also gives definition and distinction between various particles in a gas. Meaning, in a gas filled with photons, electrons and air molecules, we can set apart those particles just by knowing its spin. I'll talk about the method how eventually.

The intrinsic spin operator $\hat{S}$ is mathematically no different from the azimuthal momentum operator, in terms of how to solve for the eigenvalue. Much like angular and linear momentum, spin is conserved. Given the 1D function $\psi(x)$ in "spin space" $|s\rangle$, where s is the spin number,

$$\hat{S}_z|s\rangle = s_z|s\rangle$$

where s_z is the z-component of spin. Therefore, using the same math as before,

$$\boxed{\hat{S}_z\psi = m_s\hbar\psi} \tag{6.4}$$

where m_s is the spin magnetic number. Just like m_l having $2l+1$ values, m_s has $2s+1$ values ranging from $-s, \ldots, 0, \ldots s$.

Now I can explain how to distinguish particles based on spin. Previously, when we looked at the energy number n and the azimuthal number l, we said that these numbers are whole integers. However, for the spin number s, it can be either a whole or a half integer number. Thus, particles are defined whether they have a whole integer or a half integer spin.

Particles with whole integer spins are called "bosons." An example of a boson is the photon (the quantum particle of the electromagnetic field), which has a spin number $s = 1$, thus having $2(1)+1 = 2\,m_s$ values of $1, 0, -1$. Another example of a boson is the *graviton*, the quantum particle of gravity, which has a spin number $s = 2$ with $2(2) + 1 = 5\,m_s$ values of $\pm 2, \pm 1, 0$. Provided that bosons can have a $m_s = 0$ term, numerous bosons are free to occupy one common energy state, and still conserve the spin of the system.

Particles with half integer spins are called "fermions." An example of a fermion is the electron (a familiar particle from Vol.s II and III), which has a

6.2. INTRINSIC MOMENTUM - SPIN

spin number $s = 1/2$ with $2(1/2) + 1 = 2\, m_s$ values of $1/2$ and $-1/2$. Other spin-1/2 particles are the protons and the neutrons, but we will focus more on the electron (especially when revisiting the hydrogen atom).

Given that the electron's magnetic values are the polarized values of its spin number, it can be though that electrons have a "spin-up" and a "spin-down" orientation from the z-axis. Because electrons cannot have a $m_s = 0$ term, numerous electrons are restricted from occupying a common energy state, for it violates the conservation of spin. The only way to have $m_s = 0$ is for a coupled electron pair to have anti-parallel spins, so that $1/2 + (-1/2) = 0$.

Therefore, no third electron can occupy an energy state already housing a coupled electron pair. This is called the Pauli Exclusion Principle, named after German physicist Wolfgang Pauli.

Determining the magnitude of spin is, again, no different from the math used to find the magnitude of azimuthal angular momentum. Therefore,

$$\boxed{|\hat{S}|\psi = \hbar\sqrt{s(s+1)}\psi} \qquad (6.5)$$

where the spin magnitude is a constant for all types of particles. For a photon, $|S| = \hbar\sqrt{1(1+1)} = \hbar\sqrt{2}$; for an electron, $|S| = \hbar\sqrt{1/2(1/2+1)} = \hbar\sqrt{3/4}$; and for a graviton, $|S| = \hbar\sqrt{2(2+1)} = \hbar\sqrt{6}$.

While the azimuthal number l changes depending on the particle's energy state, the particle's spin number s is a set constant for a particular particle. Consider the Earth, which has both intrinsic and orbital angular momentum. If the Earth is not in orbit, its magnetic field still allows the Earth of have intrinsic spin. A quantum particle may have an azimuthal number $l = 0$ (given it is in $n = 1$), but its spin number s is unchanged by the azimuthal momentum.

Recall the Bohr model of a hydrogen atom (1 proton, 1 electron). The lonely electron is at the $n = 1$ energy state, meaning it has no azimuthal momentum (despite being in an orbit; this is because it has a de Broglie wavelength that encompasses the orbital path - anywhere in the wavelength there is probability of an electron being at a singular point). Yet it still has a spin number of $s = 1/2$. If a second electron joins the electron at $n = 1$, then the pairing must make a total spin $s = 0$ (which relates to $m_s = 0$) based on Pauli exclusion.

Figure 6.1: Wolfgang Pauli.

6.3 Total Angular Momentum

Just like in classical physics, the total angular momentum in quantum physics is the sum between the azimuthal momentum and the particle's spin:

$$\hat{J}\psi = \hat{L}\psi + \hat{S}\psi \tag{6.6}$$

However, if we expanded $\hat{J}$ as a direct sum between Eq.s (6.2) and (6.4), we would have a mess of values. Just like approaching $\hat{L}$ and $\hat{S}$ in their own

6.3. TOTAL ANGULAR MOMENTUM

"spaces," let us assign $\hat{J}$ a "coupled space," where $\psi(x) \to |j\rangle$. Therefore,

$$\hat{J}|j\rangle = j_z|j\rangle$$

$$\boxed{\Rightarrow \hat{J}\psi = m\hbar\psi} \qquad (6.7)$$

where $j = l + s$ is the total momentum number and m is the total magnetic number of $2j + 1$ values ranging from $-j, \ldots, 0, \ldots, j$.

Using an electron as an example, let us say it is in a hydrogen atom with 2 energy states $n = 1, 2$. At the ground state $n = 1$, the azimuthal number is $l = 0$, but $s = 1/2$ is constant. Therefore, at $n = 1$, $j = l + s = 1/2$, which is just the spin number. At the ground state, the total angular momentum assumes the role of spin, where m has $2j + 1 = 2$ values of spin-up and spin-down. However, at $n = 2$, $l = 1$ with $s = 1/2$. Therefore, $j = l + s = 3/2$, where m this time has $2j + 1 = 4$ values of $\pm 3/2, \pm 1/2$.

In the coupled space, where l and s are coupled into the total number j, we are presented with fewer values than we would in regular 1D space. For the n-2 state of the hydrogen atom, l would still be equal to 1. But instead of 4 magnetic numbers there would be 8 combinations of $m_l = \pm 1, 0$ and $m_s = \pm 1/2$. This phenomenon in 1D space is called the Zeeman effect (named after Dutch physicist Pieter Zeeman), which will be discussed more in depth when we look at the spin-dependent hydrogen atom.

Back in the coupled space, the magnitude of the total angular momentum is found by using the same math as before:

$$\boxed{|\hat{J}|\psi = \hbar\sqrt{j(j+1)}\psi} \qquad (6.8)$$

where (for the electron in the recent example) $|J| = \hbar\sqrt{3/4}$ at $n = 1$ and $|J| = \hbar\sqrt{15/4}$ at $n = 2$.

Figure 6.2: Pieter Zeeman.

Chapter 7
Rotational Hamiltonian

Consider the time-independent Schrödinger equation:

$$\frac{-\hbar^2}{2m}\nabla^2\psi + V(x)\psi = E\psi$$

Let us now draw our attention to *three*-dimensional space, where $\psi = \psi(x, y, z)$. If you are familiar with Vol.s I and II, I always simplify three dimensions into a singular, radial displacement r. Doing so requires me to consider "spherical coordinates." This means I would have to turn my Laplacian operator ∇^2 based on spherical coordinates:

$$\nabla^2 = \frac{1}{r^2}\partial_r\left(r^2\partial_r\right) + \frac{1}{r^2\sin\theta}\partial_\theta\left(\sin\theta\partial_\theta\right) + \frac{1}{r^2\sin^2\theta}\partial_\phi^2$$

where the angular terms based on θ and ϕ together form the square of the azimuthal momentum operator $L^2 = -\partial_\phi^2$.

From the Schrödinger equation in 3D space, we have the radial equation in a simplified 1D radius space:

$$\boxed{\frac{-\hbar^2}{2m}\left(\nabla_r^2 - \frac{l(l+1)}{r^2}\right)\psi + V(r)\psi = E\psi} \quad (7.1)$$

where $\psi = \psi(r, \theta, \phi)$, whose general solution is $\psi = R_n(r)Y_{lm}(\theta, \phi)$, where $Y_{lm}(\theta, \phi)$ is the spherical harmonic function:

$$Y(\theta, \phi) = P_l^m(\cos\theta)\exp(im\phi)$$

where $P_l^m(\cos\theta)$ is called the Legendre Polynomial, which depends on both azimuthal and its magnetic numbers.

7.1 The Hydrogen Atom

The last time we looked at the hydrogen atom was Vol. III, when we discussed Niels Bohr's model. From his revision we found Bohr's radius $a_0 \approx 0.5\text{Å}$ ($1\text{Å} = 10^{-10}\text{m}$) and the hydrogen binding energy to be -13.7 eV. Let us see if we can replicate these results using the Schrödinger radial equation.

A single electron and a proton in the hydrogen nucleus experience an attractive Coulomb potential:

$$V(r) = \frac{1}{4\pi\varepsilon_0} \frac{-e^2}{r}$$

which is put into Eq. (7.1) in place of $V(r)$:

$$\frac{-\hbar^2}{2m}\left(\nabla_r^2 - \frac{l(l+1)}{r^2}\right)\psi + \frac{1}{4\pi\varepsilon_0}\frac{-e^2}{r}\psi = E\psi$$

where the mass in the Schrödinger equation belongs to the electron in orbit.

Recall the resemblance between orbital motion and gyrating oscillations. The electron in orbit around the proton behaves like a harmonic oscillator, and do recall that the solution to the Schrödinger equation for an oscillator was

$$\psi(y) = \sum_n C_n y^n \exp\left(\frac{-y^2}{2}\right)$$

However, we cannot use this solution because of the added r-variable term in the rotational kinetic energy. Focusing on the radial function of the wave function $R(r)$, the hydrogen Schrödinger equation has been cleaned up:

$$R''(r) - \frac{l(l+1)}{r^2}R(r) + \frac{2m_e e^2}{4\pi\varepsilon_0 \hbar^2}\frac{1}{r}R(r) = \frac{-2m_e}{\hbar^2}ER(r)$$

and let $r = a\rho$ (a is length and ρ is dimensionless),

$$R''(\rho) + \left(-\frac{l(l+1)}{\rho^2} + \frac{2m_e e^2}{4\pi\varepsilon_0 \hbar^2}\frac{a}{\rho} + \frac{2m_e a^2}{\hbar^2}E\right)R(\rho) = 0$$

To clean up the equation further, let $a = 4\pi\varepsilon_0 \hbar^2/(m_e e^2)$ and $2m_e a^2 E/\hbar^2 = \lambda^2$. Here, λ is dimensionless. The equation is now

$$R''(\rho) + \left(\frac{-l(l+1)}{\rho^2} + \frac{2}{\rho} + \lambda^2\right)R(\rho) = 0$$

7.1. THE HYDROGEN ATOM

and the solution is nevertheless similar to the oscillator function:

$$R(\rho) = \sum_k C_k \rho^{k+l+1} \exp\left(-\rho(k+l+1)\right)$$

with the recursion relation

$$C_{k+1} = C_k \frac{\lambda + 2(k+l+1)}{(k+l+2)(k+l+1) - l(l+1)}$$

Here, the number k depends on both hydrogen state and the electron angular momentum, which are independent entities that are probabilistic. Letting $k = 0$, the recursion relation is simplified:

$$C_1 = C_0 \frac{\lambda + 2(l+1)}{(l+2)(l+1) - l(l+1)}$$

and recall that $n = l + 1$,

$$C_1 = C_0 \frac{\lambda + 2n}{2n}$$

The coefficient C_1 refers to the transition of an energy state higher or lower than the currently-occupied state. Because the probability of state transition depends on the energy, which is unknown until observation, let $C_1 = 0$, therefore forcing $\lambda + 2n = 0$. Therefore, the energy of the system can be solved:

$$\lambda^2 = \frac{2 m_e a^2}{\hbar^2} E = -4n^2$$

$$\Rightarrow E = \frac{-2n^2 \hbar^2}{m_e a^2} \tag{7.2}$$

where a was our length parameter from the Schrödinger radial equation:

$$\boxed{a = \frac{4\pi \varepsilon_0 \hbar^2}{m_e e^2} = 0.529 \times 10^{-10} \text{m}} \tag{7.3}$$

which is indeed the Bohr radius at $n = 1$. For larger n, $r = n^2 a$. With a known, our energy is

$$\boxed{E = \frac{-m_e e^4}{8\pi^2 \varepsilon_0^2 \hbar^2 n^2} = -13.606 \,\text{eV} \frac{1}{n^2}} \tag{7.4}$$

where -13.606 eV is absolutely close to the binding energy of hydrogen at $n = 1$. With our length and energy verified to be consistent with Bohr's derivation, let us determine the wave function.

Letting $k = 0$,
$$R(\rho) = C_0 \rho^{l+1} \exp(-\rho(l+1))$$
$$\Rightarrow R(r) = C_0 \left(\frac{r}{n^2 a}\right)^n \exp\left(-\frac{r}{na}\right)$$
so that the whole wave function ψ is
$$\psi(r, \theta, \phi) = C_0 \left(\frac{r}{n^2 a}\right)^n \exp\left(-\frac{r}{na}\right) Y_{lm}(\theta, \phi)$$

After normalization, the whole wave function becomes
$$\psi_{nlm}(r, \theta, \phi) = \left(\frac{r}{na}\right)^l \exp\left(-\frac{r}{na}\right) L_{n-l-1}^{2l+1}\left(\frac{2r}{na}\right) Y_{lm}(\theta, \phi) \qquad (7.5)$$

where $L_{n-l-1}^{2l+1}(2r/na)$ is called the Laguerre Polynomial. At ground state $n = 1$, the electron wave function in the 1s-hydrogen atom is

$$\boxed{\psi_{100}(r, 0, 0) = \left(\frac{1}{\pi a^3}\right)^{1/2} \exp\left(-\frac{r}{a}\right)} \qquad (7.6)$$

and the 2p-shell ($n = 2$, $l = 1$, $m = 0$) wave function is

$$\boxed{\psi_{210}(r, \theta, 0) = \left(\frac{1}{32\pi a^3}\right)^{1/2} \frac{r}{a} \exp\left(-\frac{r}{2a}\right) \cos\theta} \qquad (7.7)$$

7.2 Spin-dependent Hydrogen Atom

The spin-dependent hydrogen atom is an example of "perturbation theory," which will be discussed in depth in a future chapter. However, to put it in a short summary, a perturbation is when an exterior force jostles the system, disrupting the equilibrium of the main interaction. For our hydrogen atom (one proton, one electron), our perturbation will be an exterior magnetic potential $\hat{H}_{\text{int}} = -\vec{\mu} \cdot \vec{B}$.

In quantum electrodynamics, when charged fermions are in a region of perpendicular magnetic field, their 1/2-spins (which would either be up or down in the vacuum) become aligned in the same direction of the magnetic

7.2. SPIN-DEPENDENT HYDROGEN ATOM

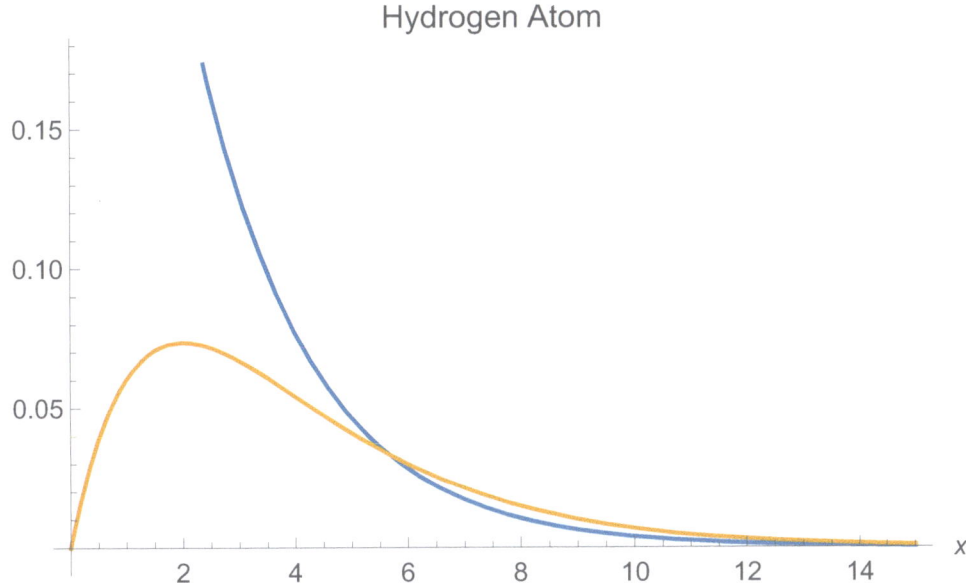

Figure 7.1: The wave functions for an azimuthal-symmetric electron in the hydrogen atom ($x = r/a$). The blue curve is the 1-s shell function and the orange curve is the 2-p function.

field. Because this forced alignment takes energy, it actually dampens the total Hamiltonian:

$$\widehat{H}_{\text{tot}} = \widehat{H}_{\text{Coulomb}} - \widehat{H}_{\text{int}} \tag{7.8}$$

where $\widehat{H}_{\text{Coulomb}}$ is the original Hamiltonian for the hydrogen atom with an energy eigenvalue of $E = -13.606/n^2$ eV.

The exterior magnetic potential becomes an operator: $\widehat{H}_{\text{int}} = -\widehat{\mu}B$, where $\widehat{\mu}$ is the magnetic moment operator:

$$\widehat{\mu}\psi = \frac{e}{2m_e c}\left(\hat{L} + 2\hat{S}\right)\psi = \frac{\hbar e}{2m_e c}(m_l + 2m_s)\psi$$

where $\hbar e/(2m_e c) = \mu_B$ is the Bohr magneton.

Therefore, the total Hamiltonian is

$$\widehat{H}_{\text{tot}}\psi = \left(-13.606\,\text{eV}\frac{1}{n^2} + \mu_B B\,(m_l + 2m_s)\right)\psi \tag{7.9}$$

with the total energy eigenvalue of

$$E_{\text{tot}} = -13.606\,\text{eV}\frac{1}{n^2} + \mu_B B\,(m_l + 2m_s) \tag{7.10}$$

At the ground state $n = 1$, $l = 0$ ($m_l = 0$) and electron spin is $s = 1/2$ with $m_s = \pm 1/2$. The total energy is therefore

$$\boxed{E_1 = -13.606\,\text{eV} \pm \mu_B B} \tag{7.11}$$

where the plus/minus also depends on the direction of the magnetic field. If there isn't any exterior field, then the total energy is the original hydrogen binding energy.

At $n = 2$, $l = 1$ ($m_l = \pm 1, 0$) and the electron spin is still $s = 1/2$ with $m_s = \pm 1/2$. This is where we have the Zeeman effect:

$$\boxed{E_2 = \frac{-13.606}{4}\,\text{eV} + \mu_B B \cdot \begin{cases} +2 & (m_l = +1, \quad m_s = +1/2) \\ +1 & (m_l = 0, \quad m_s = +1/2) \\ 0 & (m_l = \pm 1, \quad m_s = \mp 1/2) \\ -1 & (m_l = 0, \quad m_s = -1/2) \\ -2 & (m_l = -1, \quad m_s = -1/2) \end{cases}} \tag{7.12}$$

where the first column is the sum of the magnetic numbers, the second column lists the possible m_l values, and the third column lists the possible m_s values.

Part IV

Approximations

Chapter 8

Variational Method

In the past chapters on the 1D Schrödinger equation, the harmonic oscillator and the hydrogen atom; we came across certain Hamiltonians that can be solved precisely. Meaning, we solved the Schrödinger equation in order to get our wave function ψ and our energy eigenvalue E. However, as the systems become more complex, the corresponding Hamiltonians become unsolveable (at least with infinite precision).

Because the variabilities of nature are just as unpredictable as the whole of quantum mechanics, we mathematicians and theorists always need to be one step ahead. We say that mathematics is the language of nature, and thus we must assert validity towards the statement despite variabilities. The best way to do so is to approximate.

One method of approximation is the "variational method." Meaning, for bound states, the time-independent Schrödinger equation can be written as a set of variational equations. Instead of an energy eigenvalue, we have an energy *functional* that depends on an arbitrary wave function ψ:

$$\boxed{\mathrm{E}[\psi] = \int \psi^* \widehat{\mathrm{H}} \psi \, dx} \qquad (8.1)$$

Here, $\mathrm{E}[\psi]$ is the "expected energy" of the approximated Hamiltonian. This is what we expect theoretically, but all confident assurance must come from observation and / or experimentation.

However, to make the expected energy more precise, we must drive the expected energy down to the bound state. Mathematically, we take a "path derivative" with respect to a length parameter (a derivative where the path of the energy must be specific, so that any extreme variations can be chiseled away) and set it equal to zero:

$$\delta E[\psi] = \delta \left(\int \psi^* \hat{H} \psi \, dx \right) = 0 \qquad (8.2)$$

Once the derivative is taken and the length parameter of lowest energy is solved for, the length will then be put back into the expected energy to have the solution for approximated energy.

Let's assume we do not know how to solve for the harmonic oscillator. At the lowest energy state $n = 1$, the particle barely has enough energy to gyrate with translational energy. For particles that are not gyrating, they behave like a particle in a box with a symmetric Gaussian distribution:

$$\psi(x) = B \exp\left(\frac{-b^2 x^2}{2}\right)$$

where (because the wave function is a Gaussian) the normalization coefficient is $B = \sqrt{b/\sqrt{\pi}}$.

Because the Hamiltonian is for a harmonic oscillator,

$$\hat{H}\psi(x) = \frac{-\hbar^2}{2m}\psi''(x) + \frac{1}{2}m\omega^2 x^2 \psi$$

$$= \sqrt{\frac{b}{\sqrt{\pi}}} \exp\left(\frac{-b^2 x^2}{2}\right) \left[\frac{-\hbar^2}{2m}[b^4 x^2 - b^2] + \frac{1}{2}m\omega^2 x^2 \right]$$

and the energy functional can be solved to be

$$E[b] = \int \frac{b}{\sqrt{\pi}} \exp(-b^2 x^2) \left[\frac{-\hbar^2}{2m}[b^4 x^2 - b^2] + \frac{1}{2}m\omega^2 x^2 \right] dx$$

$$\Rightarrow \frac{b^4 \hbar^2 + m^2 \omega^2}{4mb^2}$$

Taking the derivative of $E[b]$ with respect to b,

$$\partial_b E[b] = \partial_b \left[\frac{b^4 \hbar^2 + m^2 \omega^2}{4mb^2} \right] = 0$$

$$\Rightarrow b = \sqrt{\frac{m\omega}{\hbar}}$$

which is the inverse of the oscillator length parameter.

Placing b into E[b],
$$E[b] \Rightarrow \frac{1}{2}\hbar\omega$$
which is exactly the energy of the harmonic oscillator (however at $n = 0$)! As we already know with $n = 1$, the precise solution for oscillator energy is $E = (3/2)\hbar\omega$, but the approximation method is pretty close to precise.

As you can see, given the physical system has an unsolveable Hamiltonian, the variational method can produce a solution of the expected energy that would be very, very close to the real solution.

8.1 Binding Energy of Helium

One main example of the variation method is the binding energy of helium (two protons, two neutrons, two electrons). Because of the added proton, the two neutrons and the additional electron, we cannot directly use the Bohr model to solve for the binding energy of neutral helium.

Let us set our two electrons at the ground state $n = 1$, which satisfies the Pauli Exclusion Principle, and let's also assume they are not so close to where they scatter away from the atom due to repulsion. Based on the fact the two electrons are at the ground state of the atom, let's assign each electron a $\psi_{100}(r, 0, 0)$ function with length parameter β:

$$\psi_e^{(1)} = \left(\frac{1}{\pi\beta^3}\right)^{1/2} \exp\left(\frac{-r}{\beta}\right)$$

$$\psi_e^{(2)} = \left(\frac{1}{\pi\beta^3}\right)^{1/2} \exp\left(\frac{-R}{\beta}\right)$$

where r and R are the displacements of the two individual electrons. The coupled electron function becomes

$$\psi_e = \psi_e^{(1)}\psi_e^{(2)} = \left(\frac{1}{\pi\beta^3}\right) \exp\left(\frac{-(r+R)}{\beta}\right) \tag{8.3}$$

The Hamiltonian of the helium atom also considers perturbation theory, this time there is no external fields disrupting the helium. We consider the kinetic energies of the individual electrons, the Coulomb potentials between the nucleus and each electron, and finally the potential between the two electrons, themselves:

$$\widehat{H} = \frac{-\hbar^2}{2m_e}\nabla_r^2 + \frac{-\hbar^2}{2m_e}\nabla_R^2 + Ze^2\left(\frac{1}{r} + \frac{1}{R}\right) + \frac{e^2}{(r+R)} \quad (8.4)$$

where $4\pi\varepsilon_0$ is scaled away, $Z = 2$ is the atomic number of helium, and

$$\widehat{H}\psi_e = \exp\left(\frac{-(r+R)}{\beta}\right)\left[\frac{-\hbar^2}{\beta^5 m_e \pi} + \frac{e^2}{\beta^3 \pi(r+R)} + \frac{Ze^2}{b^3 \pi}\left(\frac{1}{r} + \frac{1}{R}\right)\right]$$

The integral to solve for the expected energy is quite intricate and very extensive, so I will jump to the end result of the integration:

$$\mathrm{E}[\beta] = \iint \psi_e^* \widehat{H}\psi_e\, dr dR \Rightarrow \frac{-0.0253\hbar^2}{\beta^6 m_e} - \frac{0.1154 e^2}{\beta^5}$$

Solving for β at the lowest energy level,

$$\partial_\beta \mathrm{E}[\beta] = \frac{6(0.0253)\hbar^2}{\beta^7 m_e} + \frac{5(0.1154)e^2}{\beta^6} = 0$$

$$\boxed{\Rightarrow \beta = 0.2635 \frac{\hbar^2}{m_e e^2} = 0.2635 a_0} \quad (8.5)$$

which is an integer of the Bohr radius.

Considering that the radius of hydrogen is just a_0, the helium radius at ground state is smaller. This is considering that the nucleus is stronger, thus attracting each of the electrons closer. Because heavier atoms have larger radii, this is considering an abundancy in energy states, with the n-1 state ever so slightly approaching the nucleus.

Putting $\beta = 0.2635 a_0$ into the energy functional, we solve for the expected binding energy of helium:

$$\boxed{\mathrm{E}[\beta] \Rightarrow -5.5662 \frac{\hbar^2}{m_e a_0^2} = -75.7\,\mathrm{eV}} \quad (8.6)$$

which is extremely close to the real helium binding energy of -79.005 eV determined through experimentation.

For approximating binding energies, it is best to be above the real value, for bonds are stronger when the energy itself is stronger. And especially with the energies in the electron-volt scale, the approximated value is essentially the real value in the macroscopic Joule scale.

Chapter 9

Perturbation Theory

"Perturbation theory" is one other method of approximation in quantum mechanics. As we have seen with the spin-dependent hydrogen atom and the two ground state electrons in the helium atom, perturbation involves some disturbance in the system's equilibrium, to jostle the interaction into doing something other than what was initially expected.

In classical physics, an example of a perturbation is "chaos theory:" one small change in any parameter leads to something chaotic and unpredictable. The perturbation was the change in the parameter that completely made the interaction go awry. In electrodynamics, a perturbation would be when a moving charge particle enters a *magnetic* field (electric fields don't perturb charged particles, for example orbital electrons in an atom. The system here is harmonic; anything disturbing the harmony, like an exterior magnetic field, is a perturbation).

Here in quantum mechanics, let us consider a seemingly-simple collision between two particles. As I am sure we're familiar by now, each quantum particle is defined by a wave function ψ. What we may acknowledge as a simple head-on collision from a classical standpoint, particle collisions (such as scattering, repulsion, attraction, etc.) is defined by a wave interference in quantum mechanics. Meaning, the evolution of a wave is defined by how much the original wave is affected by an exterior wave.

Quantum perturbation theory is rather important to cosnider, especially going into modern quantum mechanics and quantum field theory (how each of the classical fields are described as particle interactions).

9.1 General Theory

For notation purposes, I will write my wave function $\psi(x)$ in the "interaction space," so that $\psi(x) \to |\psi\rangle$. $|\psi\rangle$ is actually no different from $\psi(x)$, for the interaction space is the 1D space. Writing the wave function in the former case would make it easier to explain perturbation theory. Note that, due to normalization, $\langle\psi|\psi\rangle = 1$.

Consider the time-independent Schrödinger equation

$$\widehat{H}|\psi\rangle = E|\psi\rangle$$

In the past, aside from spin-dependent hydrogen and neutral helium, we came across Hamiltonians that are "unperturbed," meaning unaffected by any outside sources. In perturbation notation, these Hamiltonians, wave functions and energy eigenvalues were all "zero-order:"

$$\widehat{H}^{(0)}|\psi^{(0)}\rangle = E^{(0)}|\psi^{(0)}\rangle$$

But considering a stray particle of wave function $|\phi\rangle$ comes into the system and perturbs $|\psi\rangle$, the original wave function of the first particle is a sum of higher orders:

$$|\psi\rangle = |\psi^{(0)}\rangle + |\psi^{(1)}\rangle + |\psi^{(2)}\rangle + \cdots + |\psi^{(n)}\rangle$$

where each order of $|\psi\rangle$ has a corresponding Hamiltonian:

$$\widehat{H} = \widehat{H}^{(0)} + \widehat{H}^{(1)} + \widehat{H}^{(2)} + \cdots + \widehat{H}^{(n)}$$

as well as a corresponding energy eigenvalue:

$$E = E^{(0)} + E^{(1)} + E^{(2)} + \cdots + E^{(n)}$$

Let's start simple by finding the first-order perturbation. Defining the net Hamiltonian, wave function and energy by having the highest order be 1, the Schrödinger equation looks like

$$(\widehat{H}^{(0)} + \widehat{H}^{(1)})(|\psi^{(0)}\rangle + |\psi^{(1)}\rangle) = (E^{(0)} + E^{(1)})(|\psi^{(0)}\rangle + |\psi^{(1)}\rangle)$$

which is expanded out between zero-order terms and first-order terms.

The zero-order terms come out to be the original, unperturbed equation:

$$\widehat{H}^{(0)}|\psi^{(0)}\rangle = E^{(0)}|\psi^{(0)}\rangle$$

9.1. GENERAL THEORY

while the first-order terms are written as an expansion:

$$\widehat{H}^{(0)}|\psi^{(1)}\rangle + \widehat{H}^{(1)}|\psi^{(0)}\rangle = E^{(0)}|\psi^{(1)}\rangle + E^{(1)}|\psi^{(0)}\rangle$$

Our goal now is to determine the first-order perturbation of the wave function ($|\psi^{(1)}\rangle$) in the perspective of both the original first particle function ($|\psi^{(0)}\rangle$) and the perturbing wave function of the second particle ($|\phi^{(0)}\rangle$).

Mutiplying the first-order expansion with the conjugate of the original function $\langle\psi^{(0)}|$, we will define the Hamiltonian of the first-order perturbation:

$$\langle\psi^{(0)}|\widehat{H}^{(0)}|\psi^{(1)}\rangle + \langle\psi^{(0)}|\widehat{H}^{(1)}|\psi^{(0)}\rangle = \langle\psi^{(0)}|E^{(0)}|\psi^{(1)}\rangle + \langle\psi^{(0)}|E^{(1)}|\psi^{(0)}\rangle$$

$$\boxed{\Rightarrow \langle\widehat{H}^{(1)}\rangle = E^{(1)}} \tag{9.1}$$

where $\langle\psi^{(0)}|\psi^{(1)}\rangle = 0$ and $\langle\psi^{(0)}|\psi^{(0)}\rangle = 1$.

Here, the perturbed energy of the first particle is the expected energy of the first-order Hamiltonian in the reference of the zeroth-order function, which indeed refers to the previous chapter on variational method.

This time, let's multiply the first-order expansion with the conjugate of the second particle function $\langle\phi^{(0)}|$,

$$\langle\phi^{(0)}|\widehat{H}^{(0)}|\psi^{(1)}\rangle + \langle\phi^{(0)}|\widehat{H}^{(1)}|\psi^{(0)}\rangle = \langle\phi^{(0)}|E^{(0)}|\psi^{(1)}\rangle + \langle\phi^{(0)}|E^{(1)}|\psi^{(0)}\rangle$$

$$\Rightarrow E^{(0)}_\phi \langle\phi^{(0)}|\psi^{(1)}\rangle + \langle\phi^{(0)}|\widehat{H}^{(1)}|\psi^{(0)}\rangle = E^{(0)}_\psi \langle\phi^{(0)}|\psi^{(1)}\rangle$$

where $\langle\phi^{(0)}|\psi^{(0)}\rangle = 0$ and $\langle\phi^{(0)}|\psi^{(1)}\rangle$ is defined as

$$\langle\phi^{(0)}|\psi^{(1)}\rangle = \frac{\langle\phi^{(0)}|\widehat{H}^{(1)}|\psi^{(0)}\rangle}{E^{(0)}_\psi - E^{(0)}_\phi}$$

where $E^{(0)}_\psi$ is the initial energy of the first particle and $E^{(0)}_\phi$ is the initial energy of the second particle.

Now multiplying the above expression by the original second particle function $|\phi^{(0)}\rangle$,

$$|\psi^{(1)}\rangle = |\phi^{(0)}\rangle \frac{\langle\phi^{(0)}|\widehat{H}^{(1)}|\psi^{(0)}\rangle}{E^{(0)}_\psi - E^{(0)}_\phi} \tag{9.2}$$

which defines the first-order perturbation of the first particle wave (in the interaction space). In 1D space,

$$\boxed{\psi^{(1)}(x) = \frac{\phi^{(0)}(x)}{E_\psi^{(0)} - E_\phi^{(0)}} \int \phi^{*(0)} \left(\widehat{H}^{(1)} \psi^{(0)}(x) \right) dx} \quad (9.3)$$

In essence, the first-order alteration of the reference wave function is proportional to the zero-order wave function of an interacting particle. As we consider collisions, the total momentum after a collision is completely driven by the collision itself.

9.2 Free Particle Scattering

Consider a stationary reference particle being collided and scattered by a kinetic secondary particle. In classical physics, we could consider an elastic collision and the conservation of both linear momentum and kinetic energy. In quantum mechanics, we refer to Eq. (9.3) for the scattered wave of the first particle.

For free particles, the zero-order Hamiltonian has no potential ($V(x) = 0$). However, there would be a first-order, perturbing interaction between the two particles. Let us assume our reference particle is relatively stationary, giving it a Gaussian function:

$$\psi^{(0)}(x) = \exp\left(-ax^2\right)$$

and the incoming particle have a typical sine function:

$$\phi^{(0)}(x) = \sin(bx)$$

where a and b are arbitrary constants. Let the first-order Hamiltonian be a sample collision term with an operator:

$$\widehat{H}^{(1)} = \mu f \nabla$$

where $\mu = m_1 m_2 / (m_1 + m_2)$ is the reduced mass between the two particles and f is some constant.

The zero-order psi function acting on the first-order Hamiltonian becomes

$$\widehat{H}^{(1)} \psi^{(0)}(x) = -2a\mu f x \exp\left(-ax^2\right)$$

where the integral is evaluated to be

$$-2a\mu f \int x \sin(bx) \exp(-ax^2)\, dx = -\mu f b \sqrt{\frac{\pi}{a}} \exp(-b^2/4a)$$

As for the individual initial energies, let the first particle be stationary $E_\psi^{(0)} = 0$ and the second particle have a kinetic energy $E_\phi^{(0)} = m_2 v_{2,0}^2/2$. Therefore, the first order wave function of the first particle is

$$\psi^{(1)}(x) = \frac{2m_1}{(m_1+m_2)}\frac{fb}{v_{2,0}^2}\sqrt{\frac{\pi}{a}} \exp(-b^2/4a) \sin(bx) \tag{9.4}$$

To clean the function up, let us determine the arbitrary constants. a and b are place holders that neutralize the neighboring units, i.e. $[b] = 1/m$ and $[a] = m^{-2}$. But f is coupled with mass and the del operator (with units of inverse-meter) to form the units of energy, i.e. $[f] = m^3/s^2$.

One can speculate that $f = \bar{v}_f^2/b$, where $\bar{v}_f$ is the average velocity between the scattered particles after collision. So,

$$\boxed{\psi^{(1)}(x) = \frac{2m_1}{(m_1+m_2)}\frac{\bar{v}_f^2}{v_{2,0}^2}\sqrt{\frac{\pi}{a}} \exp(-b^2/4a) \sin(bx)} \tag{9.5}$$

As for the first-order energy of the reference particle, it is the expectation of the first order Hamiltonian with respect to the zero-order psi function:

$$\boxed{E^{(1)} = \int \psi^{*(0)} \hat{H}^{(1)} \psi^{(0)} \Rightarrow \mu \bar{v}_f^2} \tag{9.6}$$

9.3 Electron Repulsion

Let us consider another scenario: electron repulsion. Two free electrons move horizontally to clash head-on, only to be richochetted by their electric repulsion. Both electrons are initially moving, let's make it simple by saying simultaneously at the same rate:

$$\psi^{(0)}(x) = \sin(\kappa x)$$

$$\phi^{(0)}(x) = \cos(\kappa x)$$

where κ is some constant.

The first-order Hamiltonian would be a repulsion potential between the two electrons:
$$\widehat{H}^{(1)} = ke^2 \nabla$$
where $k = 1/4\pi\varepsilon_0$ is the Coulomb constant from classical electrostatics.

The zero-order psi function acting on the first-order Hamiltonian becomes
$$\widehat{H}^{(1)} \psi^{(0)}(x) = ke^2 \kappa \cos(\kappa x)$$
where the integral is evaluated to be
$$ke^2\kappa \int \cos^2(\kappa x) dx = ke^2\kappa(n\pi)$$
where n is a whole integer depending of the order of perturbation. For first order, $n = 1$.

As for the initial energies, both electrons were moving simultaneously and identically. However, provided that Eq. (9.3) has the energy difference in the denominator, identical energies would result to a zero, making the whole function be a calculation error. However, let us consider one electron to be in a rest energy frame, and another in a relativistic frame, i.e. $E_\psi^{(0)} = \gamma m_e c^2$ and $E_\phi^{(0)} = m_e c^2$.

Therefore, the first order wave function of the first particle is
$$\psi^{(1)}(x) \approx \left(\frac{e^2}{4\varepsilon_0 m_e c^2}\right) \frac{\kappa}{\gamma} \cos(\kappa x)$$

$$\boxed{\Rightarrow \alpha \frac{\pi \hbar}{m_e c} \frac{\kappa}{\gamma} \cos(\kappa x)} \qquad (9.7)$$

where $\alpha = 1/137$ is the fine structure constant of quantum electrodynamics and $h/m_e c$ is the Compton wavelength of an electron, which is analagous to an x-ray / gamma ray.

As for the first-order energy of the reference particle, it is the expectation of the first order Hamiltonian with respect to the zero-order psi function:

$$\boxed{E^{(1)} = \int \psi^{*(0)} \widehat{H}^{(1)} \psi^{(0)} \Rightarrow \frac{e^2 \kappa}{8\pi\varepsilon_0}} \qquad (9.8)$$

or half of the electric repulsion potential.

Part V

Relativistic Quantum Mechanics

Chapter 10

Special Relativity and Quantum Mechanics

The past chapters focused on *non-relativistic* quantum mechanics, which was the developing research from the early 1920's. As we have seen, non-relativistic quantum mechanics was no simple walk in the park as I may have shown it to be. Remember, I'm the one writing this - so you don't have to.

It was a radical shift to look at particles as waves instead of concentrated matter, and thus the conceptual baggage weighed in on the complexity of quantum mechanics. Now we're making the baggage slightly heavier with *relativistic* quantum mechanics.

As we recall from Vol. III, Albert Einstein approached his 1905 theory of relativity by imagining how a ray of light perceives the world around it. From this daydream, Einstein would conceive $E = mc^2$, the energy-momentum theory for the massless particles and the photoelectric effect for scattering electrons and photons.

And as special relativity came across a speed bump when considering constant acceleration, Eintein would derive general relativity ten years later in 1915. And with it, the theory of black holes from Karl Schwarzschild and the notion of gravitational waves. As we discuss relativistic quantum mechanics derived in the late 1920's and on, it serves as a bridge between Einstein's relativity and the quantum mechanics from the early 1920's.

10.1 Klein-Gordon Equation

The Klein-Gordon equation, named after physicists Oskar Klein of Sweden and Walter Gordon of Germany, is an equation derived in 1926. It looks at Einstein's energy-momentum theorem in quantum operator form. From which, it becomes a wave equation for a certain m_s-1 particle (the photon).

Recall the energy-momentum theorem from Vol. III:

$$E^2 = p^2c^2 + m^2c^4$$

where relativistic energy is seen as the sum of the object's momentum and its rest energy. If momentum is zero ($v = 0$), then the energy reduces down to the rest energy. Therefore, in the quantum sense, the particle has a time-dependent wave function based on $i\hbar \partial_t \psi = mc^2 \psi$.

The Klein-Gordon equation requires to look at non-zero momentum (non-zero velocity). Therefore, using the operators $\widehat{E} = i\hbar \partial_t$ and $\widehat{p} = -i\hbar \nabla$ for energy and momentum, respectively,

$$-\hbar^2 \partial_t^2 \psi(x,t) = -\hbar^2 c^2 \nabla^2 \psi(x,t) + m^2 c^4 \psi(x,t)$$

which can be revised as in terms of the d'Alembert wave operator $g^{\mu\nu}\partial_\mu \partial_\nu = \partial_t^2/c^2 - \nabla^2$ from Vol. III:

$$\hbar^2 \left(\nabla^2 - \frac{1}{c^2} \partial_t^2 \right) \psi(x,t) = m^2 c^2 \psi(x,t)$$

From this, the Klein-Gordon equation is derived to be

$$\boxed{-g^{\mu\nu} \partial_\mu \partial_\nu \psi(x,t) = \frac{m^2 c^2}{\hbar^2} \psi(x,t)} \tag{10.1}$$

where the right hand side is proportional to the inverse-square of the particle's Comption wavelength $\lambda = h/(mc)$. Note that $\hbar = h/(2\pi)$. For a photon with a particular wavelength,

$$-g^{\mu\nu} \partial_\mu \partial_\nu \psi(x,t) = \frac{4\pi^2}{\lambda^2} \psi(x,t) \tag{10.2}$$

where $2\pi/\lambda = k$ is the wave number.

In the quantum-mechanical sense, a photon has the spin value $s = 1$, which gives the particle $m_s = +1, 0, -1$ (spin up, no spin, spin down). But since the equation is composed of squares, the m_s values are optimized to be positive: $m_s = 1, 0$ (either present or not).

For the Klein-Gordon equation is an "inhomogeneous wave equation" (the d'Alembertian is non-zero; a zero d'Alemberian is the basic wave equation from Vol.s I and II), the optimized spin value is $m_s = 1$. Therefore, the Klein-Gordon equation is an equation for photons. Such solutions for Eq. (10.2) shall be the same for the classical EM waves, provided some scaling corrections.

10.2 Dirac Equation

The Dirac equation, named after Englishman Paul Adrian Maurice Dirac, is an equation derived in 1928. Dirac approached Einstein's energy-momentum theorem a bit differently from Klein and Gordon. This equation considers the electron's spin number of $s = 1/2$, and thus requires a spin-dependent direction.

Dirac original looked at the original Klein-Gordon equation, and thought to describe energy in the first-order power: $E = \sqrt{p^2c^2 + m^2c^4}$, or

$$\sqrt{\left(\nabla^2 - \frac{1}{c^2}\partial_t^2\right)}\psi(x,t) = \frac{mc}{\hbar}\psi(x,t)$$

To incorporate the electron spin relativistically, Dirac would consider that the d'Alembertian operator would be a result of a square of the following operator:

$$\left(\nabla^2 - \frac{1}{c^2}\partial_t^2\right) = \left(A\nabla + \frac{i}{c}B\partial_t\right)\left(A\nabla + \frac{i}{c}B\partial_t\right)$$

where A and B are square matrices (an array of entries / elements arranged in n rows and n columns) of unknown elements. However, they are normalized and anti-commutative, such that $AB + BA = 0$ and $A^2 = B^2 = 1$. Replacing the d'Alembertian as the above, its square root become one of the terms:

$$\left(A\nabla + \frac{i}{c}B\partial_t\right)\psi(x,t) = \frac{mc}{\hbar}\psi(x,t)$$

$$\Rightarrow \left[\hbar\left(A\nabla + \frac{i}{c}B\partial_t\right) - mc\right]\psi(x,t) = 0$$

Recall that A and B are matrices, and that they somehow incorporate the electron spin relativistically. Let us intoduce a 4-dimensional spin matrix

Chapter 10. Special Relativity and Quantum Mechanics

Figure 10.1: Paul Adrian Maurice Dirac.

γ^μ, where $\mu = 0, 1, 2, 3$ for the four dimensions of specetime. Let us expand the del operator to consider 3D space,

$$\left[\hbar\left(A_1\partial_x + A_2\partial_y + A_3\partial_z + \frac{i}{c}B_0\partial_t\right) - mc\right]\psi(x,t) = 0$$

where I denote the matrix B to be B_0. Therefore, I can determine what my

10.3. A BRIEF INTRO TO QUANTUM FIELD THEORY

A and B matrices are:
$$\boxed{B_0 = \gamma^0} \qquad (10.3)$$
for the time parameter, because $x_0 = ct$, and
$$\boxed{A_n = i\gamma^n, \text{ where } n = 1, 2, 3} \qquad (10.4)$$
for the spatial parameters, because $x_1 = x$, $x_2 = y$ and $x_3 = z$.

Putting A_n and B_0 in the equation,
$$\left[i\hbar\left(\gamma_0 \frac{1}{c}\partial_t + \gamma_1\partial_x + \gamma_2\partial_y + \gamma_3\partial_z\right) - mc\right]\psi(x,t) = 0$$
we see that the γ^n components and coordinate deriviatives define a trace of the spin matrix: $\gamma^\mu \partial_\mu$, where ∂_μ is the contravarient gradient $\partial/\partial x^\mu$ from Vol. III. Thus, we generalize the above to define the Dirac equation:
$$\boxed{(i\hbar\gamma^\mu \partial_\mu - mc)\psi(x,t) = 0} \qquad (10.5)$$

Because the Dirac equation is spin-dependent for an electron, meaning each element of the γ^μ matrix is based on the up and down orientations, the mass in the equation is for the electron: $m = m_e$.

The Dirac equation defines the motions of a kinetic particle particularly based on its momentum. Consider the momentum operator in 3D space, $\widehat{p} = i\hbar\nabla$. For 4D spacetime, the del operator becomes the contravarient gradient: $\nabla \to \partial_\mu$, and thus the Dirac equation can be defined by relativistic momentum as
$$\gamma^\mu \widehat{p}_\mu \psi = m_e c \psi$$

10.3 A Brief Intro to Quantum Field Theory

Quantum field theory is one of the frontiers of theoretical physics, which we attempt to describe the same physics of classical fields in the quantum scale. As each of the the four fundamental fields in nature (electromagnetism, the strong and weak nuclear forces, and gravitation) has their classical framework, either purely or semi-classical, quantum field theory considers a "field-particle duality" between the particle and the field it harnesses.

In classical field theory, the electric, magnetic and gravitational fields bring rise to a particular type of force, by which a test object is influenced

Chapter 10. Special Relativity and Quantum Mechanics

by a source object. For such fields, the strength of the field is dependent on the source via its mass, its charge and/or its motions.

In *quantum* field theory, the essence is still there, but the mathematics must conform to the quantum paradigm. Previously, we looked at discreet particles (discreet meaning being set to one possible location within a defined region) as a probability wave function, the larger peaks indicate the larger likelihood of its location. In field theory, the particle is no longer discreet: it is continuous (set to all possible locations within the region).

Consider the hydrogen atom: a proton and an electron. In the classical theory, a proton has radially inward field lines that illustrate the electric attraction. An electron would have to follow the arrow of the field to carry out the attraction. Since electrons have themselves a wave function based on the likelihood of their location in the atom, *relativistically*-moving electrons traced along the electric field "move" their wave functions to trace out the electric field, thus creating a string-like "spectrum function:"

$$\psi \to \Psi$$

These spectrum functions apply to all particles (not just for electrons), and for every particle, there is an anti-particle (meaning a near-identical copy-cat of their particle partner, just with one crucial thing that set them apart).

As we look at the Dirac equation, Dirac also tells us that each electron occupies an "electron state." The number of these states correspond to the abundancy of particles present. However, the occupancy of electron states inversely affects the occupancy of the anti-particle "positron" states.

Provided a nearly-equal amount of occupied electron and positron states, the electrons and the positrons may exist upon the same region of space to interact in a particle/anti-particle collision. These collisions are fundamental to quantum field theory, for they give the reason why a classical (or semi-classical) phenomenon occurs. Once an electron collides with the positron elastically, it emits a photon, conserving charge ($-e + e = 0$) and spin ($1/2 + 1/2 = 1$). This is called pair-production.

Chapter 11

Gravity and Quantum Mechanics

The pillars of modern physics today are Einstein's general relativity (the physics of the very large) and quantum mechanics (the physics of the very small). To conceive a theory of everything, general relativity and quantum mechanics must satisfy one another. Recall the Einstein equations for general relativity from Vol. III:

$$G_{\mu\nu} = \frac{8\pi G}{c^4} T_{\mu\nu}$$

where the "Einstein tensor" $G_{\mu\nu}$ can be rewritten as the negative Laplacian $-\partial_\mu \partial_\nu / 2$. We used this approach when looking at gravitational waves, as finding the trace of the tensor leads to the d'Alembert wave operator. Just as the Hamiltonian was redefined as a wave equation to become the Schrödinger equation, we can use general relativity to propose a quantum wave equation:

$$-g^{\mu\nu} \partial_\mu \partial_\nu \psi = \frac{16\pi G}{c^4} g^{\mu\nu} T_{\mu\nu} \psi$$

Here, the metric is linearized: $g^{\mu\nu} = \eta^{\mu\nu} + h^{\mu\nu}$: $\eta^{\mu\nu}$ is the canonical metric and $h^{\mu\nu}$ is the perturbation metric.

This is where quantum gravity comes to a crossroads: how to approach the metric, effectively taking the direction of either geometry or interaction. This also affects the neccessary approach for the stress-energy tensor, either taking its trace via the canonical metric or the perturbation metric. As we will discuss, taking the interaction picture is the route of string theory, and thus the geometry picture is the route of loop quantum gravity.

It was Richard Feynman who once said, "If you cannot describe a concept of physics in laymans terms, you do not know it yourself." As I am writing this, and thinking about how to progress from here on, the more Feynman is proven to be right.

The subject of quantum gravity is a realm of mystery and uncertainty. Over the past half a century, the theories of string theory and loop quantum gravity were drafted to combine quantum mechanics with Albert Einstein's general theory of relativity. As spot on they might be on their own, the wider the gap between these two theories becomes.

11.1 String Theory

String theory is one candidate for a quantum field theory of gravitation, which looks at gravity as an interaction. Thus, for our open-ended Einstein equation, we use $h^{\mu\nu}$ for the metric. As for the stress-energy tensor, we want to show that the energy density the tensor traces is defined as a particle field under a path of least action. As the gravitational field in the classical framework is a conservative field, we would want to preserve "conservatism" in the quantum field definition.

So, we use the Lagrangian $\mathcal{L}$ (in this case, the Lagrangian density), and we define the stress-energy tensor as

$$T_{\mu\nu} = h_{\mu\nu}\mathcal{L} - 2\frac{\delta\mathcal{L}}{\delta h^{\mu\nu}}$$

Using the perturbation metric and the stress-energy tensor defined as a Lagrangian, we have

$$-h^{\mu\nu}\partial_\mu\partial_\nu\psi = \frac{16\pi G}{c^4}h^{\mu\nu}\left[h_{\mu\nu}\mathcal{L} - 2\frac{\delta\mathcal{L}}{\delta h^{\mu\nu}}\right]\psi$$
$$= \frac{32\pi G}{c^4}\mathcal{L}\psi$$

with $[h^{\mu\nu}h_{\mu\nu} - 2(\delta h^{\mu\nu}/\delta h^{\mu\nu})] = 2$. Thus, we define the Lagrangian from the Einstein equation:

$$\boxed{\mathcal{L}\psi = -\frac{c^4}{32\pi G}h^{\mu\nu}\partial_\mu\partial_\nu\psi} \qquad (11.1)$$

where $c^4/(32\pi G) = \kappa^2$ is a gravitational coupling constant typically used to describe graviton interactions (a graviton being the spin-2 quantum particle

11.1. STRING THEORY

of gravity). We can relate this Lagrangian with the Einstein-Hilbert action:

$$S = \frac{c^4}{16\pi G} \int R\sqrt{-\det(g_{\mu\nu})} d^4x$$

by recalling $g^{\mu\nu}\partial_\mu\partial_\nu = -2R$ (in our case, $g^{\mu\nu} = h^{\mu\nu}$). From the Einstein-Hilbert action, we define the action expression for particle-fields:

$$\begin{aligned} S &= -\frac{c^4}{32\pi G} \int h^{\mu\nu}\partial_\mu\partial_\nu \sqrt{-\det(g_{\mu\nu})} d^4x \\ &= \int \mathcal{L} dt \end{aligned} \quad (11.2)$$

where $dt = \sqrt{-\det(g_{\mu\nu})} d^4x$.

11.1.1 Particles as Strings

As Arthur Compton connected wavelengths to corresponding particle masses, we can approach particles as closed loops (or strings) which are vibrating. These vibrations intensify proportional the the particle's mass, charge, spin, or any of its quantum characteristics. We use Bohr's wavelength equivalence:

$$n\lambda = 2\pi \Delta s$$

where λ is the Compton wavelength and Δs is the string metric. Uniquely, the vibrations upon a particle string can be described as the action upon that string. Thus, we define $n\hbar = -\mathcal{L}\Delta t$ for the action integral:

$$\begin{aligned} -\mathcal{L}\Delta t &= mc\Delta s \\ \rightarrow -\mathcal{L} dt &= mc\, ds \end{aligned}$$

so that we have

$$S = \int \mathcal{L} dt = -mc \int ds$$

As we're describing vibrating strings, the rest energy of the particle mc^2 can be equivalent to a characteristic string tension T_0, so that $mc^2 = T_0 s$. This changes the action to be

$$S = -\frac{T_0}{c} \int dA$$

where $dA = d(s^2)$ is the area element of the "world sheet" traced by the string metric. A world sheet is a two-dimensional cylinder-like surface traced by the particle string with a spatial parameter σ and a time parameter τ. In 4D spacetime, the string would have the "worldline" function:

$$X(\sigma,\tau) = X^0(\sigma,\tau)\mathbf{e}_0 + X^1(\sigma,\tau)\mathbf{e}_1 + X^2(\sigma,\tau)\mathbf{e}_2 + X^3(\sigma,\tau)\mathbf{e}_3$$

each sub-function of the worldline function represents each of the dimensions the particle is occupying. Applying Stokes' theorem on the integration of the world sheet, the area of the world sheet is revised as the surface element:

$$dA = d\sigma c d\tau \sqrt{-\det(g_{ab})}$$

where the metric matrix of the world sheet g_{ab} is made of the spatial and time diffrenetials of the worldine function:

$$\det(g_{ab}) = (\partial_\tau X)^2 (\partial_\sigma X)^2 - (\partial_\tau X \cdot \partial_\sigma X)^2$$

Therefore, the relativistic action of the particle string is defined as

$$\boxed{S = -T_0 \int d\sigma d\tau \sqrt{(\partial_\tau X \cdot \partial_\sigma X)^2 - (\partial_\tau X)^2 (\partial_\sigma X)^2}} \quad (11.3)$$

which is called the "Nambu-Goto Action." To make the string action more length-dependent, Nambu and Goto introduced a "slope paramater" α', which rewrites the action as

$$S = -\frac{1}{2\pi\hbar c\alpha'} \int d\sigma d\tau \sqrt{(\partial_\tau X \cdot \partial_\sigma X)^2 - (\partial_\tau X)^2 (\partial_\sigma X)^2} \quad (11.4)$$

where string tension is defined as

$$T_0 = \frac{E}{s} = \frac{1}{2\pi\hbar c\alpha'}$$

Note that the energy is the rest energy of a particle with relativistc string vibrations, and this displacement parameter s outlines the string metric. Should one take a linear string, and wrap it to where the end points touch, this circular loop traces out the outer casings of a physical particle, making $s = 2\pi l_s$, where l_s is the string length. Because particles are vibrating nodes, they would have rest energy similar to photonic energy: $E = \hbar c/l_s$, where the string length l_s acts as wavelength. Therefore,

$$\frac{\hbar c}{2\pi l_s^2} = \frac{1}{2\pi\hbar c\alpha'}$$

$$\boxed{\Rightarrow l_s = \hbar c \sqrt{\alpha'}} \quad (11.5)$$

11.1. STRING THEORY

Because particles have a unique quantum property called spin, the spin of a particle directly affects the length of the string; the larger the spin, the longer the length. Since α' is related to the inverse square of energy unitwise, to incorporate spin (generally angular momentum) the slope parameter is defined as

$$\alpha' = \frac{J}{\hbar E^2} \tag{11.6}$$

where $J = \hbar\sqrt{j(j+1)}$ is quantum angular momentum and E is the particle energy:

$$\Rightarrow \alpha' = \frac{1}{m^2 c^4}\sqrt{j(j+1)} = \frac{\lambda^2}{\hbar^2 c^2}\sqrt{j(j+1)}$$

This make the string tension be defined as follows:

$$T_0 = \frac{1}{2\pi\alpha'\hbar c} = \frac{m^2 c^3}{2\pi\hbar\sqrt{j(j+1)}} \tag{11.7}$$

and the string length be defined in terms of the particle wavelength:

$$\boxed{l_s = \lambda\,(j(j+1))^{1/4}} \tag{11.8}$$

which is not always the Compton wavelength of a rest particle.

Here, the string length is dependent on the angular momentum of the particle:

$$\mathrm{L}|\psi\rangle = l\,(j(j+1))^{1/4}\,|\psi\rangle$$

where L is the "length operator" and l is the corresponding length eigenvalue.

11.1.2 Dimensions as Membranes

Since particles are depicted as vibrations along a string, and to which a world sheet, this world sheet can be unraveled into a vibrating spacetime membrane, or a "brane", whose geometrical warpings are the physical entity of a particle. The depth of curvature is described by the particle's energy amplitude and vibrational frequency.

Every particle has its own field, and the "stacking" of various particle fields upon one another outlines the quantum interaction of these particles in a given region of spacetime. Provided the interactions of various particles in a region of spacetime, these stacked particle fields would look quite like a vibrating region of local "quantum" spacetime.

Should an auxillary, rigid metric rest at the equilibrium of the particle fields, each hump surfacing from the flat field would look like emerging worldsheets. These worldsheets can be reduced down to strings, whose end points are attached to the flat field. Such strings are called "open strings," named as such due to their open ends fastened to the flat field. Open strings are the physical material that "hold together" a set of 4D spacetime, meaning the everyday particles we are constantly in contact with.

Suppose there is a second "brane," acting like the highest energy state above the ground state. Certain particle fields can reach to energy amplitudes that can penetrate the next higher brane. As it requires energy to transition into energy states, it requires energy to "leak" into the next highest brane. Such particles that can do so effortlessly are called "closed strings," opposite to the "open string" particles such as electrons and photons. A notorious example of a closed string is the graviton.

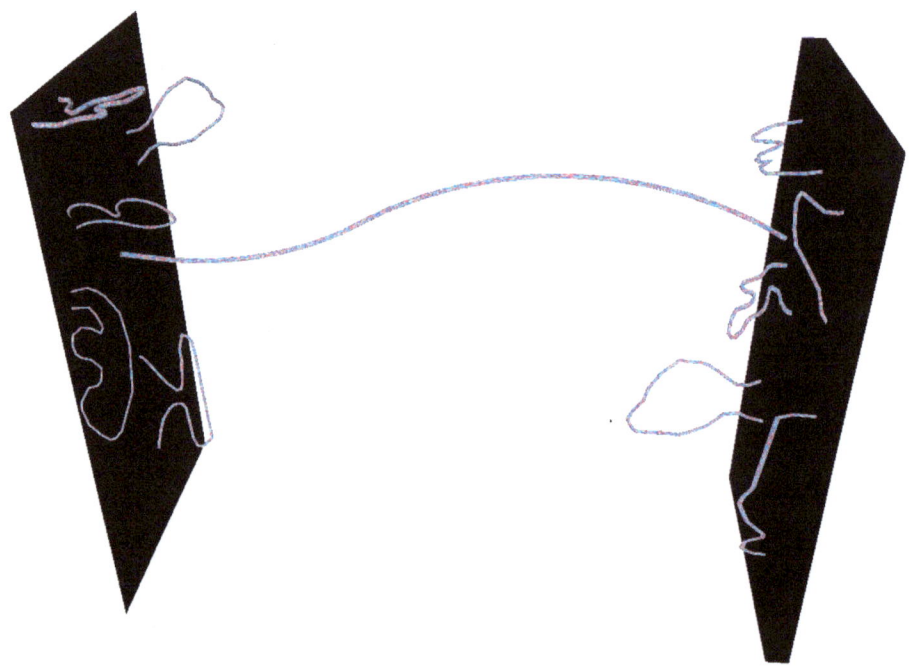

Figure 11.1: Two 4D spacetime membranes connected by a string.

Like much of quantum mechanics, "leakage" is a spontaneous, stochastic event that can happen without any warning. However, we can focus on the "ground state brane", and what is going on canonically. This continuous "fizzing" of quantum fields emerging from the rigid metric is essential to the

11.2 Loop Quantum Gravity

Loop quantum gravity (LQG) is the second candidate theory to gravitational quantum field theory. Instead of looking at gravity as an interaction, LQG looks at gravity as canonical geometry. Thus, for our open-ended Einstein equation, we use $\eta^{\mu\nu}$ for the metric. Approaching the stress-energy tensor is going to be intricate to say the least.

When we describe canonical spacetime, the local region of spacetime is unbent, flat. However, due to the Heisenberg uncertainty priciple, a quantum spacetime could not possibly be at a stand-still. There would be quantum curvatures that, from the macroscopic perspective, is flat (much like viewing ocean waves from outer space versus at sea level). So, we can look at our quantum spacetime as a grid-like lattice while we define our trace $\eta^{\mu\nu}T_{\mu\nu}$ as two terms:

$$\eta^{\mu\nu}T_{\mu\nu} = \frac{\hbar c}{16\pi}(q_{ab}q_{cd} - \frac{1}{2}q_{ac}q_{bd})\delta^{ac}\delta^{bd} - \frac{c^4}{16\pi G}\det(q)R[q]$$

where q is a generalized variable. Here, I explicitly write the Latin indicies for each of the 4 coordinates. Each term is for two fundamental components of a "Planck-scale" region of spacetime.

The first term depicts the "matter" that composes quantum spacetime: an elementary "chunk" that cannot be extracted as a free particle. With its form similar to that of a simple harmonic oscillator, these "spacetime chunks" act like vibrating lattice points that help refrain the spacetime from collapsing into a singularity.

The second term depicts a "force mediator" between two sample "chunks", as gluons are for quarks inside protons and neutrons. Keeping up with the lattice analogy, these are the connections between two lattice points that keep the lattice intact. We can intuitively say, due to the $c^2/(16\pi G)$ factor, a connection between the spacetime matter is a graviton field.

We have Einstein's equation revised as

$$-\eta^{\mu\nu}\partial_\mu\partial_\nu\psi = \left[\frac{G\hbar}{c^3}(q_{ab}q_{cd} - \frac{1}{2}q_{ac}q_{bd})\delta^{ac}\delta^{bd} - \det(q)R[q]\right]\psi$$

where $G\hbar/c^3 = l_P^2$ is the square of the Planck length, $l_P = 1.6 \times 10^{-35}$ m. As it depicts canonical spacetime, there is no change in the metric: $\eta^{\mu\nu}\partial_\mu\partial_\nu = 0$. Thus, we have

$$\left[\frac{G\hbar}{c^3}(q_{ab}q_{cd} - \frac{1}{2}q_{ac}q_{bd})\delta^{ac}\delta^{bd} - \det(q)R[q]\right]\psi = 0 \qquad (11.9)$$

This is the Wheeler-DeWitt equation. In simple terms, it provides a Hamiltonian constraint on canonical spacetime:

$$\widehat{\mathrm{H}}\Psi_h = 0 \qquad (11.10)$$

meaning these Planck-sized curvatures destruct into a flat surface:

$$\widehat{\mathrm{H}}\Psi_h = \sum_{i=0} \widehat{\mathrm{H}}^{(i)}\psi_{(i)} = 0$$

Each Hamiltonian in the summation resembles the warping of smaller "sub-metrics." This network of sub-metrics is based on the ADM Formulism, which states that spacetime is a set of sub-spaces with its own non-negative Hamiltonian. This is essentially the Casimir effect in mathematical writing.

Figure 11.2: A spin foam in loop quantum gravity.

11.2. LOOP QUANTUM GRAVITY

11.2.1 The Loops in LQG

The warping of sub-metrics via the ADM Formulism is very similar to the "stacking" of particle fields in string theory. These quantum ripples upon the sub-metrics resemble a quantum "foam," each foam has a corresponding length operator:

$$\widehat{L}\Psi_h = \frac{l_P}{2\kappa'}\sqrt{\sqrt{j(j+1)}}\,\Psi_h$$

where the length eigenvalue is proportional to the Planck length. The factor $\kappa' = 1/\sqrt{32\pi}$ is the graviton coupling without the physical constants. Here, j depicts the spin number of these connections, which can be half-integer for flat spacetime or whole-integer for bent spacetime.

Unlike string theory, where the length eigenvalue would represent the Compton wavelength of a vibrating particle string, the length eigenvalue represents a unit length that separates a quantum chunk from another. If looking at Albert Einstein's bendable wire mesh spacetime continuum from general relativity, $\sim 5l_P$ is a length from one grid point to another.

Meaning, roughly five Planck lengths is a side length of a grid square in the quantum lattice. Each of these grid squares are actually the chunks of spacetime interwoven and fastened by the connections into a "loop." The whole lattice forms a "spin network," and a local region of spin networks form the foam.

Like worldsheets in string theory, these spin networks have a surface area. Here, the area of a spin network is done by coupling two length operators into an area operator (much like the scaling theory of area as $A = L^2$):

$$\widehat{A}\Psi_h = \left(\widehat{L}\cdot\widehat{L}\right)\Psi_h = \frac{l_P^2}{4\kappa'^2}\sqrt{j(j+1)}\,\Psi_h \qquad (11.11)$$

from that we define the area eigenvalue for a spin network:

$$\boxed{A = 8\pi l_P^2 \sqrt{j(j+1)}} \qquad (11.12)$$

In the context of loop quantum gravity, Einstein's spacetime is a fizzing, foaming surface. Each bump and hill comes and goes sporatically. Since each "foam bubble" appears and disappears, this is similar to the Casimir effect, how particle-antiparticle pairings appear and collide to preserve the quantum stability of spacetime.

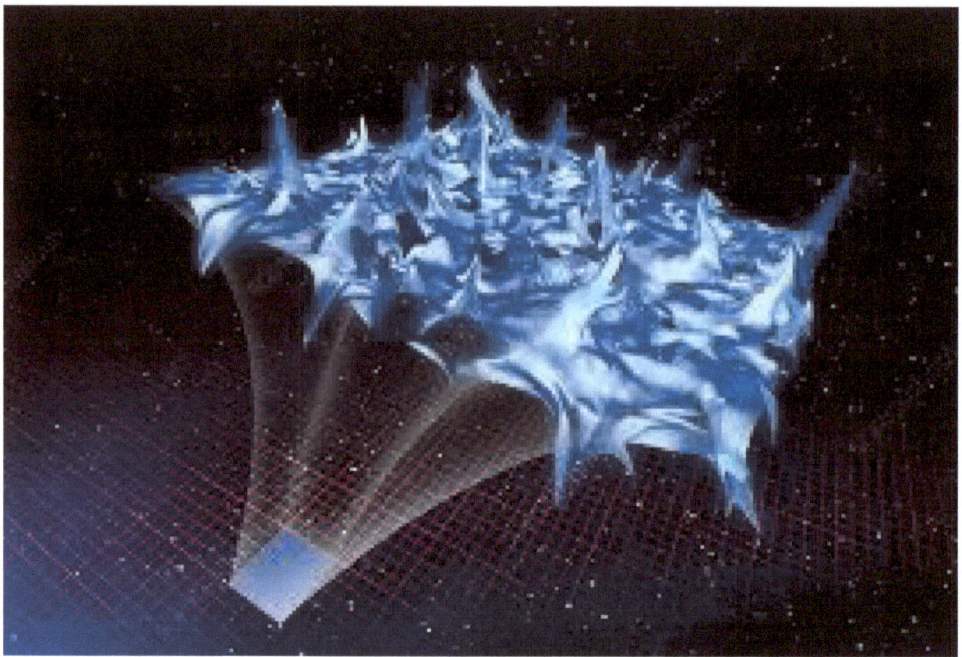

Figure 11.3: A representation of a spin network in loop quantum gravity.

To that extension, LQG preserves the validity of Hawking radiation. But direct experimentation is far from tangible. To test that quantum spacetime is indeed a foaming surface, gamma ray bursts were tested to see that the time of arrival is increased due to a form of friction from the spin networks. The accuracy of time arrival is however spot on to the calculation.

11.3 Quantum Particle of Gravity

The year was 1934; two Soviet physicists named Dmitrii Blokhintsev and F.M. Gal'perin hypothesized that there has to be a quantum particle that harnesses the classical gravitational interaction between two celestial bodies. It would have to be massless and chargeless – just like the photon for the electromagnetic interaction. But instead of being a spin-1 boson (as photons are, whose spin correlates to the rank-one electric and magnetic field vectors), it has to be a spin-2 boson, correlating to the rank-two stress-energy tensor from Einstein's general relativity. Blokhintsev and Gal'perin were theorizing the graviton.

In the 1950s, it was becoming clear in Einstein's gravity theory that any object with a rest energy (even massless stuff) can warp the local spacetime

11.3. QUANTUM PARTICLE OF GRAVITY

around it. Of course, the context here is a classical object constructed by localized radiation, but the scale is rather left ambiguous. These classical orbs of energy were called "gravitational geons," or simply "geons," which can be thought as a classical particle of gravity (independent of all sorts of quantum weirdness). However, with a paradigm shift from modern physics to quantum physics, the question of the classical geon finally made its way towards the quantum graviton.

In order to have a successful quantum field theory for gravitation, the graviton must be the field boson of the gravitational field. It also needs to be in wave-particle duality with gravitational waves. Should the gravitational field be quantized into a current of gravitons, it must be thought that celestial bodies induce this leakage of gravitons just as they induce classical fields. As general relativity discusses gravity as the warping of spacetime due to mass and energy, gravitons must show some connection to geometry as well as interaction. And most importantly, gravitons are attracted to astronomical masses and free-falling objects to simulate gravitation in the quantum scale.

11.3.1 Spin of a Graviton

It was mentioned that the graviton has to be a spin-2 field boson. This is a novel proof that verifies that the graviton spin is indeed $s = 2$. Due to the wave-particle duality between gravitons and gravitational waves, let us consider a vacuum wave in relation to quantum spin. Recall that the operator for the z-component of spin is

$$\widehat{S}_z \Psi = m_s \Psi$$

where $\hbar = 1$. The spin projection number m_s takes up values that range from $+s$ to $-s$. For instance, the graviton has five projection values of $m_s = \pm 2, \pm 1, 0$. A particle with prominent spin has thus a spin projection value at either of the extreme values of $m_s = \pm s$.

From Vol. III, two vacuum waves construct into a time-independent wave that can exist over all space. The completely constructed wave amplifies the singular gravitational wave, which shall be the graviton particle with prominent spin projection. Since it represents a graviton, the amplified wave is symmetric to the singular wave:

$$\sqrt{2} \exp[ikr] = m_s \frac{1}{\sqrt{2}} \exp[ikr]$$

where each wave function is time symmetric. Therefore,

$$\boxed{m_s = 2} \tag{11.13}$$

which is by definition the positive integer of s, verifying that $s = 2$.

The origin of the graviton spin being $s = 2$ comes from the rank-two stress-energy tensor. Because the tensor is the source of Ricci curvature via gravitational induction, the stress-energy tensor can also be the source of purturbations such as gravitational waves. As previously reviewed in Vol. III, celestial bodies with torsion do produce negligible gravitational waves that could likely be considered "gravitonic".

11.3.2 Gravitational Wave-Particle Duality

Because gravitons are only acknowledged to have a wave-particle duality with gravitational waves, yet not proven to have so, this section is rather investigative and naive. Proceed with caution when reading this. Consider the classical gravitational wave equation from Weyl (tidal) curvatures:

$$\left(\frac{1}{c^2}\partial_t^2 - \nabla^2\right) h(x,t) = \frac{-16\pi G \epsilon_W}{c^4} h(x,t)$$

where $h(x,t) = N\psi_h(x,t)$ (N is the number of gravitons in a classical wave), and $(\partial_t^2/c^2 - \nabla^2) = h^{\mu\nu}\partial_\mu\partial_\nu$.

We want to make the claim that classical gravitational waves contain N gravitons, each with its own "quantum" energy. Supposing the Weyl energy density has a classical-quantum correspondance, we write $\epsilon_W = N\mathcal{H}$. This single "quantum" energy, $\mathcal{H}$, shall be the Hamiltonian energy $\mathcal{H} = \mathcal{L} - p_\mu u^\mu$, which includes the Lagrangian from string theory and a massless relativistic energy $p_\mu u^\mu = U_0$. For now, we ignore spin.

Thus, the classical wave equation becomes a quantum wave function:

$$-h^{\mu\nu}\partial_\mu\partial_\nu\psi_h = \frac{16\pi G}{c^4}\left[-\frac{c^4}{32\pi G}h^{\mu\nu}\partial_\mu\partial_\nu - U_0\right]\psi_h$$
$$= -\frac{1}{2}h^{\mu\nu}\partial_\mu\partial_\nu\psi_h - \frac{16\pi G}{c^4}U_0\psi_h$$

where we can define a plausible graviton wave equation:

$$\boxed{h^{\mu\nu}\partial_\mu\partial_\nu\psi_h = \frac{1}{\kappa^2}U_0\psi_h} \tag{11.14}$$

11.3. QUANTUM PARTICLE OF GRAVITY

This spin-less wave equation is equal but opposite to the Lagrangian.

Alternatively, to consider spin, we refer to the Wheeler-DeWitt equation as a sum of destructing subspace Hamiltonians (reaching up to third order):

$$\widehat{H}\Psi_h = \widehat{H}^0 \psi_h^{(0)} + \widehat{H}^1 \psi_h^{(1)} + \widehat{H}^2 \psi_h^{(2)} + \widehat{H}^3 \psi_h^{(3)} = 0$$

Since the subspace Hamiltonians in the sum are non-zero, one of which may resemble a gravitational wave in the vacuum. Bringing back the ocean analogy, the waves only take place along the surface layers of the water; the deepest layers remain unaffected. So, the quantum curvatures for an upper-layer Hamiltonian can be considered as a perturbed metric, as if it were from a force-mediating graviton.

Therefore, the Hamiltonian of the uppermost spacetime layer represents an energetic perturbation (a gravitational wave/a wave graviton), presenting a free wave kinetic term:

$$\widehat{H}^3 \psi_h^{(3)} = m_s \frac{-\hbar^2}{2m_g} g^{\mu\nu} \partial_\mu \partial_\nu \psi_h^{(3)}$$

where $m_g \approx 10^{-58}$ kg is the mass of the graviton by means of the Compton wavelength $\lambda_C \approx 10^{16}$ m. Due to the smallness of m_g, one can still call the particle massless. The term m_s is the spin projection value, which is the key aspect to spin-dependency.

The above equation perserves the Wheeler-DeWitt equation, $\widehat{H}^3 \psi_h^{(3)} = 0$, for the spin projection $m_s = 0$. Meaning, this spin-dependent kinetic term has a Hamiltonian constraint.

A graviton with a non-zero mass has four degrees of freedom. In terms of spin-dependency, these degrees of freedom are the non-zero projections $m_s = \pm 2$ and ± 1. Yet, massless gravitons have two degrees of freedom. How can we have an equation that works for both gravitons?

Mathematically, the absolute value of the projections restricts the two negative values to be positive, thus presenting two degrees of freedom. In other words, the following (speculative) equation is interchangeable for both massive and massless gravitons:

$$\boxed{\widehat{H}\psi_h = |m_s| \frac{-\hbar^2}{2m_g} g^{\mu\nu} \partial_\mu \partial_\nu \psi_h} \quad (11.15)$$

Unlike the Dirac equation, the spin-dependent relativistic equation for electrons, this spin-dependent relativistic equation for gravitons dismisses the spinor tensor γ^μ. This is because γ^μ is a matrix whose main diagonal is the "Pauli matrices," the spin matrices that specifically satisfy spin-1/2 fermions, such as electrons. Because gravitons are spin-2 bosons, the Pauli matrices cannot be considered, and with them the spinor tensor.

Only the future may tell about the future of quantum gravity, let alone *The Theory of Physics*.

Part VI

Problems

Chapter 12

Exercise Problems

Provided are ten problems that are all based on the material covered in quantum mechanics. Each problem has parts, which requires the reader to refer to the past chapters. Questions with an exclamation mark (!) are challenge problems, which provides helpful tips to consider while solving the problem.

Schrödinger Equation

1. Consider the solution for a free particle.
    a) If the velocity of the particle was $v = p/m = \hbar k/m$, how would the wave function look after the revision?
    b) Allowing $k = \pi/L$, where L is the length of the all-space region, would the free particle solution behave like a particle in an infinite well of large L (Do not assume $L = \infty$)?
    c) Letting x/L become a constant n since $L >>> x$ ($0 < n < 1$), use a graphing calulator / website / program or the unit circle to map the revised function for $n = 0.25$, $n = 0.5$, $n = 0.75$. How do these three functions compare with each other?

2. (!) We have a quantum particle of mass m under the gravitational influence of a celestial planet of mass M.
    a) Set up the Schrödinger equation if the potential of the system is $V(x) = GmM/x$. The wave function would be solved by this equation as

$$\psi(a) = a^2 \exp[-(n\pi a)^2]$$

where $a = x/R$ and R is the farthest extent to the planet's gravitational field.

b) Letting $0 \leq a \leq 1$ (so $0 \leq x \leq R$), graph the wave function within the region of a for quantum states $n = 1$, 2 and 3.

c) For the Earth with radius $x = 6.371 \times 10^6$ m, let the location of the wave function's peak be where the planet radius is marked, letting the origin be the center of the planet.

What would be the extent of the Earth's gravitational field R for the states $n = 1$, 2 and 3? If the location is not changing for increased states, then what is?

3. (!) Suppose a supermassive black hole can be treated as a symmetric linear well due to Hawking radiation. A radiation particle has a mass of $m = \hbar c/(4\pi GM)$, which depends on black hole mass. Along the surface (at the Schwarzschild radius $r_S = 2GM/c^2$), the constant force is the surface gravitational force: $F = \hbar c/(4\pi r_S^2)$, and the energy eigenvalue is the Hawking radiation energy: $E = \hbar c^3/(8\pi GM)$.

a) Find the specific expressions for the length scaling a, energy parameter ϵ, and the displacement $a\epsilon$. How does the obtained ϵ relate to the energy state n? How does the obtained $a\epsilon$ relate to the ratio E/F, meaning what is the significance of the "work applied"?

b) With the obtained a, ϵ, and $a\epsilon$, what is the normalization coefficient A_0? Does it depend on the black hole radius r_S? What is the final form of the wave function for one "Hawking particle"?

c) The energy E, temperature $T = E/k_B$, and surface force F of the system are inversely related to the black hole mass M. The larger the mass, the smaller the quantities; the smaller the mass, the larger the quantities. As for the wave function, what happens to its form when the mass is large (use $M = 10$ or 50) and when it is small (use $M = 1$ or 0.1)? How does it relate to the energy/temperature/surface force of the system?

Simple Harmonic Oscillator

4. In the energy basis, consider a potential well with states $n = 1$, 2 and 3.

a) For an oscillator at the ground state $n = 1$, what is the energy-basis Hamiltonian of the particle as it ascends to the n-3 state (Use the creation operator acting on the ground state function $|1\rangle$ as it reaches to the n-3 function $\langle 3|$; recall that $\langle i|j\rangle = \delta_{ij} = 1$ if $i = j$ or 0 if $i \neq j$).

b) If that same oscillator (that is now at $n = 3$) descends to the n-2 state, what would be the Hamiltonian now (Use the annihilation operator)?

c) If the ground state oscillator's net motion between parts (a) and (b) ended up at the n-2 state, how would the oscillator's *displacement* $\langle 2|a^\dagger|1\rangle$

compare to its distance (answers of (a) and (b) added together)?

Rotational Hamiltonian

5. Consider the gravitational Schrödinger equation from problem 2, now using the Schrödinger radial equation.

   a) Find the gravitational Bohr radius R at the ground state $n = 1$ by referring to the gravitational attraction between a proton and an electron with orbital speed $(v = \sqrt{Gm_P/R})$ and quantum angular momentum $n\hbar = m_e v R$.

   How does this compare to the radius of the observable universe $R_U = 8.8 \times 10^{26}$ m? If the calculated R is larger than R_U, then the universe has at most an n-2 energy state. If not $(R < R_U)$, then the universe is naturally at the ground state.

   b) Based on the calculated R and its relationship to R_U, what is the azimuthal number $l = n - 1$?

   c) With the knowledge obtained from parts (a) and (b), would the system between a particle and a planet change for the radial equation, or would it be the same as the 1D version?

6. For the spin-dependent hydrogen atom at the ground state $n = 1$, if the Bohr magneton is $\mu_B = 5.788 \times 10^{-5}$ eV/T and the Earth's magnetic field is 3.05×10^{-5} T,

   a) what would be the total hydrogen energy if the field was pointing down? What if pointing up?

   b) Hydrogen is a diatomic molecule, meaning two hydrogen atoms must be coupled in order to be chemically stable. If the binding energy were to double in value, while the magnetic potential of the Earth remained the same, what is the total *diatomic* hydrogen energy at the ground state?

Variational Method

7. (!) Let us look at the simple harmonic oscillator system, however where each oscillating particle has a charge q.

   a) At the ground state, an oscillator as the ansatz function of
   $$\psi(x) = bx \exp[-(bx)^2/2]$$
   If the Hamiltonian is
   $$\widehat{H} = \frac{-\hbar^2}{2m}\nabla^2 + \frac{1}{2}m\omega^2 x^2 + qfx^2$$

find $\widehat{H}\psi$.

b) The expected energy is given as

$$E[b] = \frac{3b\hbar^2\sqrt{\pi}}{8m} + \frac{3qf\sqrt{\pi}}{4b^3} + \frac{3m\omega^2\sqrt{\pi}}{8b^3}$$

Solve for b by setting $\partial_b E[b] = 0$.

c) With the b parameter solved in part (b), what is the expected energy of the charged harmonic oscillator?

Perturbation Theory

8. (!) The spin-2 graviton is in wave-particle duality with gravitational waves, and harnesses the gravitational force for celestial sources. Suppose a gravitational wave is about to come into contact with a planet.

a) If the graviton in the gravitational field has the function

$$\psi^{(0)}(x) = x^2 \exp[-x^2/2]$$

and another graviton in the gravitational wave has the function

$$\phi^{(0)}(x) = x \exp[-x^2/2]$$

with a perturbing Hamiltonian

$$\widehat{H}^{(1)} = m_g \frac{G\hbar}{c} \nabla^2$$

solve for $\widehat{H}^{(1)}\psi^{(0)}$.

b) Now solve for

$$\int \phi^{*(0)}(x) \widehat{H}^{(1)} \psi^{(0)}(x) dx$$

c) If $E_\psi - E_\phi = m_g c^2$, what is $\psi^{(1)}$, assuming the field graviton is scattered by the wave graviton? Does the field graviton become a gravitational wave if it is scattered? If so, how is $\psi^{(1)}$ different from $\phi^{(0)}$?

Relativistic Quantum Mechanics

9. Consider the Klein-Gordon Equation for a light particle with wavelength λ and angular frequency ω. If the ansatz of the wave function is $\psi = \sin(kx)\exp[-i\omega t]$ (a classical EM wave) and letting $\lambda = 2\pi/k$, is the

equation satisfied?

10. (!) Refer to the plausible spin-dependent Hamiltonian for the graviton.
   a) Suppose the Hamiltonian has additional energy terms:

   $$|m_s|\frac{-\hbar^2}{2m_g}g^{\mu\nu}\partial_\mu\partial_\nu\psi_h = i\hbar c\, \mathbf{e}_\mu \partial^\mu \psi_h - E_0\psi_h$$

   ($\mathbf{e}_\mu$ is a "vierbein" unit vector), so that (after some assumptions) the wave function is solved to be

   $$\psi_h(r,t) = A_0 J_0\left(\frac{kr}{\sqrt{|m_s|}}\right)e^{-\omega t}$$

   If normalization took place over all space $r \in (-\infty, \infty)$ and all time $t \in [0, \infty)$, what is the coefficient A_0 (Hint: $J_0(x) \simeq \sin(x)/x$)? What happens to the form of the normalized wave function under spin projections $m_s = \pm 2$ and ± 1?

   b) For the normalized wave function from part (a), plot ψ_h as a function of space, using time as an adjustable parameter. As time progresses, what happens to the wave function? Does this demonstrate a vital principle in quantum mechanics?

Part VII

Acknowledgements

Chapter 13

About the Author

Born in Natrona Heights, Pennsylvania, in 1997, Noah Matthew MacKay was raised in eastern North Carolina since 2000. In 2005 he was diagnosed with Aspergers Syndrome at the age of eight.

MacKay started a hobby of writing fiction novels in 2013 (at the age of fifteen) and keeps up with it ever since. As a novelist, MacKay wrote five books from his fantasy series titled *Age of War* between 2016 and 2020. Since 2015, Noah MacKay also writes poetry in the German language. His noteworthy poetry publications are in the following collections: a "best-of" collection of his 2015-2019 poetry and autobiographical account *Einfach ich* (Simply Me) and *Monster* (Monster), both in 2020; and the latest collection *Asyl* (Asylum) in 2021.

In addition to writing novels and poetry, MacKay wrote four first-edition volumes of the mathematical physics series *The Theory of Physics* in 2020, *The Theory of Gravity* in 2021, and the English and German versions of *Quantum Particles of Gravity* – also in 2021. Alongside his physics books, MacKay wrote the English and German versions of the book *Germany in the Twentieth Century*.

In May 2020, Noah M. MacKay graduated magnum cum laude from East Carolina University with a B.Sc. in physics and a B.A. in German language and literature. And in May 2022, he graduated with a M.Sc. in physics, also from East Carolina University. His master's thesis was on the shear viscosity of quark-gluon plasma. Academic publications include a journal article based on his master's thesis and a co-authored encyclopedia article on Weimar Republic-era poet and satirist Erich Mühsam.

Figure 13.1: Noah Matthew MacKay

Chapter 14

Thanks

Ich möchte als Danksagung erwähnen:

Die Familien MacKay und Morrison, deren Liebe, Glückwünschen und Gebete mir sehr beeinflussen, die Ausdauer zu haben, die Physik und die deutsche Sprache als Hauptfächer zu studieren und die Bachelorabschlüsse darauf zu schaffen.

Herr Professor Doktor John Kenney, der mir die Willigkeit gab, die Physik zu übertreffen.

Herr Professor Doktor Michael Dingfelder, der mir in beiden wissenschaftlichen Untersuchungen sehr half: meinem Selbststudium zur Wellen-Teilchen-Dualität der Gravitation und meiner Bachelorarbeit zur Streuungstheorien geladener Teilchen.

I would like to thank:

The MacKay and Morrison Families, who gave me their love and prayers for me to achieve two bachelor's degrees in physics and in German language.

Dr. John Kenney, who taught my first ever university physics course, who said to me "never bullshit an expert," and who gave me the inspiration to excel further into physics.

Dr. Michael Dingfelder, who taught my classical mechanics and quantum mechanics courses, supervised my independent study and my senior thesis projects, and further inspired me to pursue theoretical physics.

Wenn es einen leidenschaftlichen Gedanken gibt, den man nah beim Herzen hält, kann man den durch Kunst und Wissenschaft in der einfachsten Art ausdrücken.